I0479191

BASIC STEPS IN
GEOMETRICAL CONSTRUCTION

SOLOMON ARUOTURE

BASIC STEPS IN GEOMETRICAL CONSTRUCTION

Copyright © 2022 Solomon Aruoture

All rights reserved. No part of this book may be reproduced or used in any manner by electronic or mechanical means including information storage and retrieval systems without written permission of the copyright owner except for the use of quotations in a book review.

Published by:

Ten Over Ten Publishers

tenovertenpublishers@gmail.com

ACKNOWLEDGEMENT

Nothing in life is ever successful without the collective efforts of several gifted persons who are willing to network; submit their talents, experiences and passions for a common goal. A book like this was not written unaided. This book lends from numerous individuals' thoughts, efforts, ideas, knowledge and experiences.

First among them is my wife, my ride or die – Omoefe Solomon. I am eternally grateful for your patience and understanding during my endless late-night writing. To my beautiful angels – Anna and Annabelle, thanks for your understanding during the period of writing this book.

I would like to use this opportunity to specially thank National Examination Council (NECO), The Lagos State Examination Board, West Africa Examination Council (WAEC) and University Tertiary Matriculation Examination (UTME) for the opportunity to use their past questions for the benefits of the students and teachers.

Special appreciation to my mentors, Dr. Odumosu Mary Olukemi, Ezekiel Ojo, Prof. M.O. Ibrahim, and R. A. Jimoh, your teachings and counsel have immensely helped a great deal in the success of this publication.

FOREWORD

Basic Steps in Geometric Construction is a book useful in the study of constructions which is an aspect of mathematics that students dread most in secondary school mathematics. The book has been carefully written in a simplified, easy to follow, step by step and easy to understand language. The author provided numerous examples and guides to solve some basic construction in mathematics. The users therefore need little or no assistance to understand the concepts.

The book which deals with construction of lines, angles, triangles and quadrilaterals is full of learning activities for thorough understanding of the content. The book will be tremendously useful for both junior and senior secondary schools and others who may want to write external examinations including Joint Admission and Matriculation Board.

I strongly recommend Basic Steps in Geometrical Construction to all in Junior and Senior secondary school who wants to pass mathematics with ease and all lovers of mathematics.

DR. ODUMOSU, Mary Olukemi
Chief Lecturer
Department of Mathematics
University of Education
Formerly: Adeniran Ogunsanya College of Education
Lagos, Nigeria.

CONTENTS

CHAPTER ONE

Mathematical Instruments and Their Use

The study of points, lines, curves, angles, shapes and their properties are called Geometry. This aspect of mathematics is purely practical so, to achieve this, we need to make use of certain tools called geometrical tools (mathematical tools or instruments). Some of the most common geometrical tools we use today can be seen in "the set of mathematical instruments" - math-set.

A complete mathematical instrument contains the following:

A metal compasses

A metal divider

Pencils (17cm and 9cm)

A clear plastic ruler (15cm)

A clear plastic Set square 60^0 and 45^0

180^0 protractor

A pencil sharpener

An eraser

Alphabet and number stencil

Lab instrument stencil

Useful fact sheet

Let us now look at some of the most used geometrical instruments.

RULER

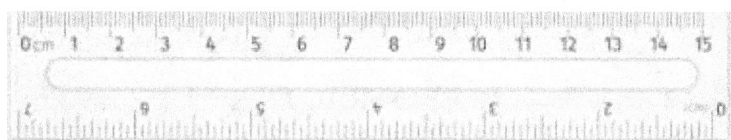

A ruler is a geometrical instrument used for constructing straight lines and also for measuring the length of a line segment. The ruler is also called *straightedge*. It is graduated into centimeters and inches. The intervals or marks on the ruler is called the *Hash Mark*.

HOW TO MEASURE OBJECTS USING A RULER

1. Place the zero-hash mark of the ruler on one end of the line or object you want to measure.
2. Properly align the object you are measuring along the edge of the ruler.
3. Read the measurement on the hash mark of the ruler, which is at the other side of the object ends.
4. Then read between the zero-hash mark and the final hash mark reading on the ruler. The difference between the hash marks reading give the exact length of the object you are measuring.

COMPASSES

The compass is a V-shaped tool that holds a pencil on one side and a steel pointer on the other side. The distance between the pencil and the steel pointer of the compass is adjustable and it's called the *radius*. It is used to draw arcs, circles, angles, and mark equal lengths. The compass can also be used to find the midpoint of a shape, line and mark intersections.

PROTRACTOR

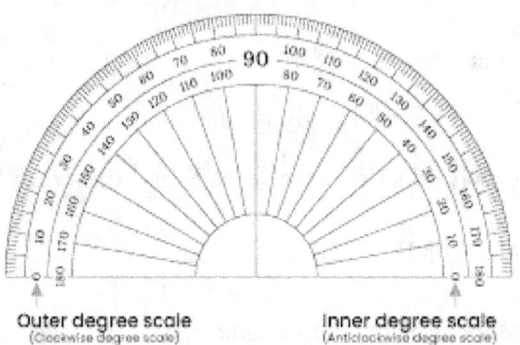

The protractor is a semi- circular instrument used for measuring and drawing angles ranging from 0^0 to 180^0. The protractor has two set of markings from 0^0 to 180^0 from left to right and from right to left. The inner and outer angles reading of the protractor will add up to 180^0. The inner reading is from right to left while the outer reading is from left to right.

It is important to note that in using the protractor; If the angle to be measured is at the left side of the protractor, we will have to use the outer reading provided in the protractor but if the angle to be measured is at the right side of the protractor, we will have to use the inner reading provided in the protractor.

HOW TO USE THE PROTRACTOR

Step 1: Line up the vertex of the angle with the dot or centre marker of the protractor

Step 2: If the angle to be measured is on the right, line up one side of the angle with the inner number scale reading 0^0 on the protractor. If the angle to be measured is on the left, line up one side of the angle with the outer number scale reading 0^0 on the protractor.

Step 3: Read the protractor to see where the other side of the angle crosses the number scale

SET SQUARES

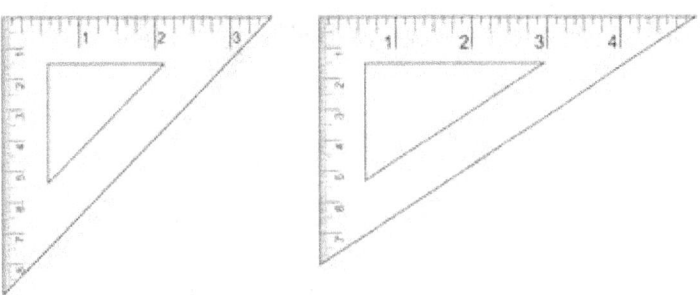

The set squares are plastic triangular instruments with some portion between them removed. It is mainly used for drawing angles 30^0, 45^0, 60^0 and 90^0 and drawing parallel and perpendicular lines. There are two kinds of set squares available in the set mathematical instruments which are 45^0 and 60^0 an both have right angle (90 degrees) in them.

DIVIDER

A divider is an instrument that looks like a compass which is used for measuring line segment, transferring or marking off distance. The divider consists of two straight adjustable legs ending in sharp point that is joined together.

PENCILS

The pencil is the most common instrument in geometry. The set of mathematical instruments comes with two pencils of size 17cm and 9cm respectively. The longer pencil is used for drawing lines while the shorter one is always used with the compass for drawing arcs, circles, angle and so on.

ERASER

An eraser is an instrument used erase unwanted lines or mistakes made with a pencil.

PENCIL SHARPENER

This is an instrument used for making the pencil's writing point very sharp so as to ensure accuracy of point and line segments.

Exercise 1:

Identify and write out the functions of the following mathematical instruments:

1.

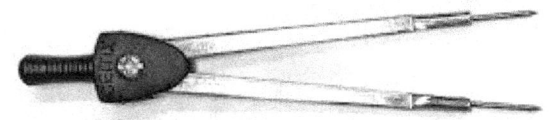

2.

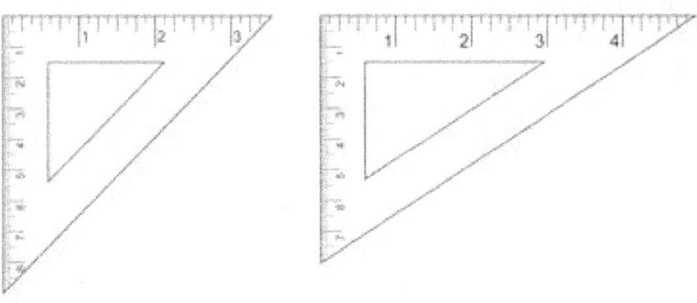

3.

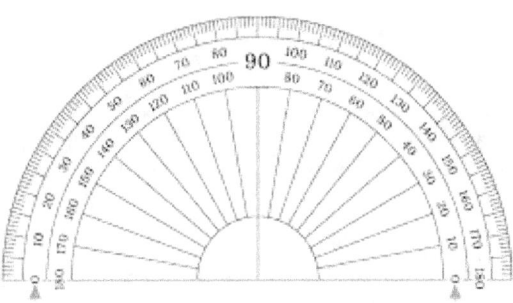

5.

6.

7.

CHAPTER TWO

Important Facts on Geometrical Construction

It is important for student to note the following before constructing:

1. Make sure your pencils are very sharp because using a blunt pencil may give you inaccurate result.

2. When drawing lines, ensure you draw a line that has no width.

3. Avoid shading or painting lines or arcs when constructing.

4. Make sure you avoid using the pencil on your compass to draws or join points together so as not to tamper with the radius of your compass

5. Ensure your eraser is clean and does not stain paper.

6. Make sure your compass is working well and not loose so as to arrive at a correct and neat construction.

7. Make sure your ruler has a good straightedge and not damaged.

8. Make it a habit to ensure that great care is taken while drawing a line through a point or two.

9. Always make a rough or free hand sketch before making an actual construction.

10. Make it a habit to always draw arcs with long radius for easy intersection. Arcs should not be thick.

11. Do not erase the construction lines, they will show the teacher or examiner how the lines or shapes were constructed.

CHAPTER THREE

Construction of Lines

Definition of Line

A line is the distance between two points. The line may be straight or curve. A line segment is denoted with a bar on top of both alphabets representing the points. $\overline{AB}$ represent line segment between point A and Point B while $|AB|$ represent the length of the line segment AB.

How To Construct a Line of Any Given Length

How to draw a straight line of length 6cm.

Step 1: Draw a line and mark a point A on the line

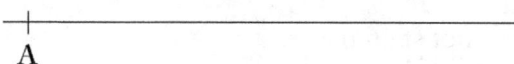

Step 2: With a pair of compasses make a radius (this is the distance between the pencil and the steel pointer of the compass) of 6cm

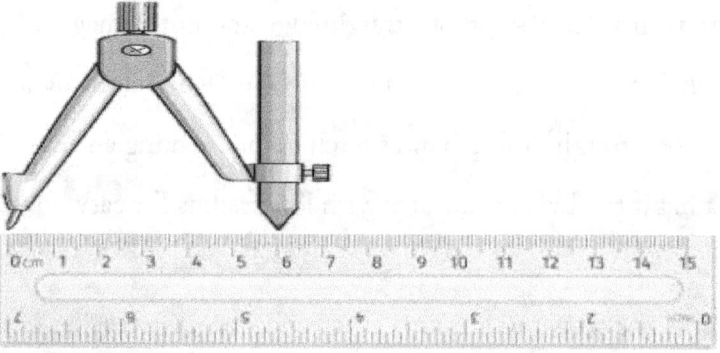

Step 3: With A as the centre, draw an arc to cut the line and call the point where it cut the line B

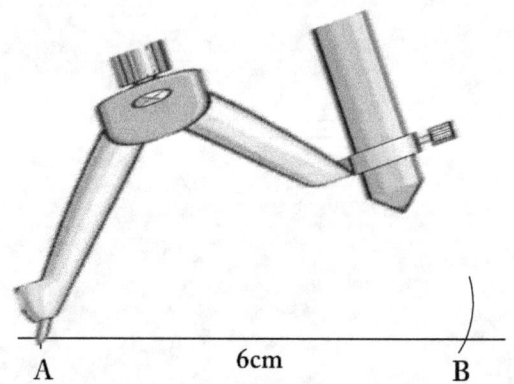

Step 4: $\overline{AB}$ is the required line, $|AB|$ = 6cm

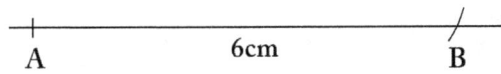

Exercise

Construct the following:

1. $|AB|$ = 5.1cm

2. $|XY|$ = 13.5cm

3. $|PQ|$ = 10.3cm

4. $|CD|$ = 7.4cm

Measure and record the length of the following lines;

5. _____

= _____ cm

6. _____

= _____ cm

7. _____

= _____ cm

8. _____

= _____ cm

Parallel Lines

Definition of Parallel Lines

Parallel lines are two or more lines that lie side by side each other which can never meet or cross each other no matter their length. The distance between the lines remains the same at any point.

How To Construct Parallel Lines.

Step 1: Draw line **AB** and mark point **C** above the line **AB**

Step 2: With point **C** as the centre, draw an arc to cut line **AB** at **D**

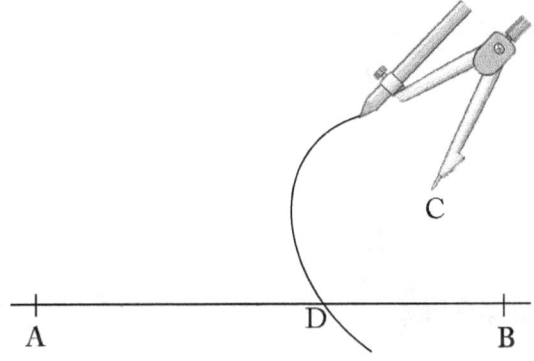

Step 3: With the same radius as above, and with point **D** as the centre draw an arc to cut line **AB** at **E**

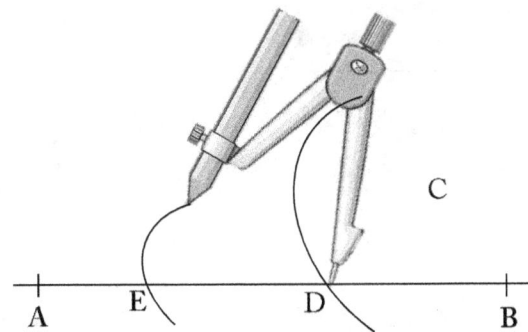

Step 4: With **E** as the centre, draw an arc to cut the arc **D** at **F**

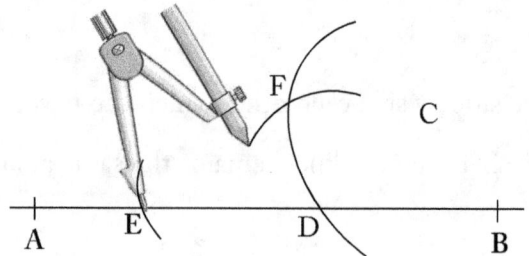

Step 5: Draw a straight line to join point **C** and **F**. $\overline{MN}$ is parallel $\overline{AB}$ mathematically it can be written as

$$\overline{MN} \parallel \overline{AB}$$

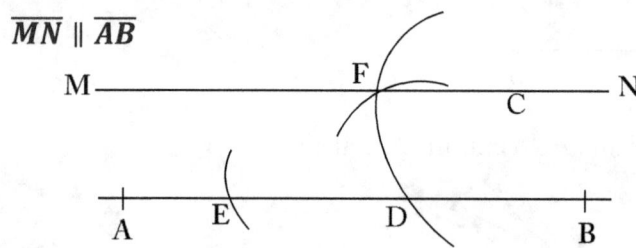

ALTERNATIVELY:

Step 1: Draw $\overline{AB}$ and mark a point C above the line

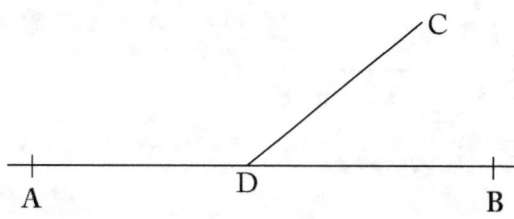

Step 2: Join point C to $\overline{AB}$ at point D

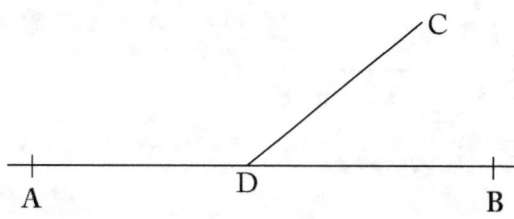

Step 3: With D as the centre, draw an arc to cut $\overline{CD}$ at point E and $\overline{AB}$ at point F respectively

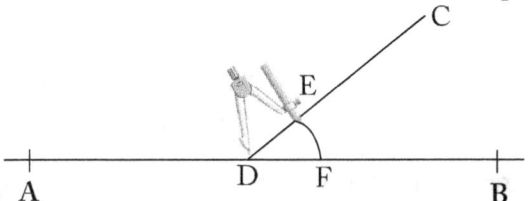

Step 4: With C as the centre, draw an arc above point C to cut the line at point G.

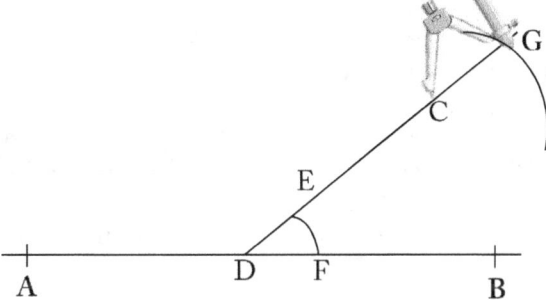

Step 5: With G as the centre, draw an arc to cut the previous arc at point H

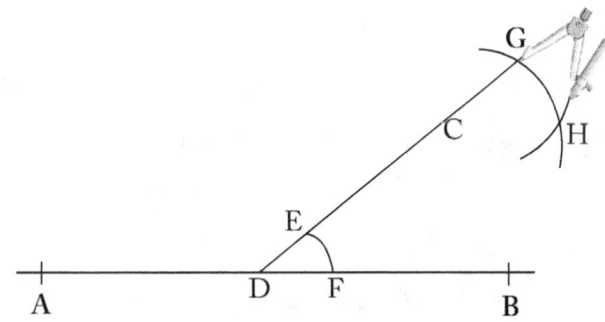

Step 6: Use ruler to join point C and H extend the line segment to $\overline{IJ}$

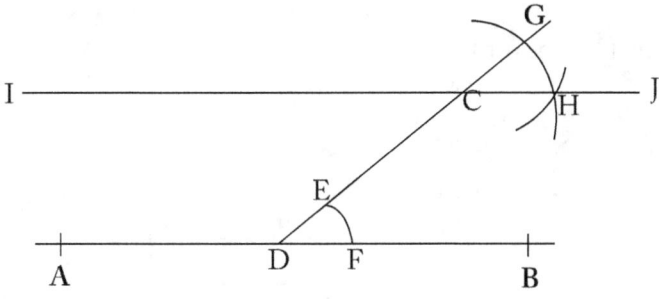

Step 7: The $\overline{IJ}$ is the required line parallel to $\overline{AB}$

Perpendicular Lines

Definition of Perpendicular Lines

A straight line which meets another straight line at right angle (90^0) is called a perpendicular. This implies that the angle between the two straight lines (one horizontal and one vertical) must be 90^0.

How To Construct Perpendicular Lines

Step 1: Draw a straight line and mark point C at the centre.

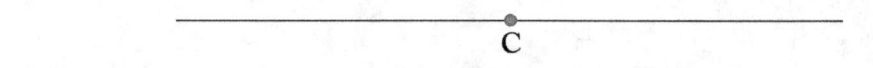

Step 2: With C as the centre, draw arcs of any radius on both sides of point C and call it A and B.

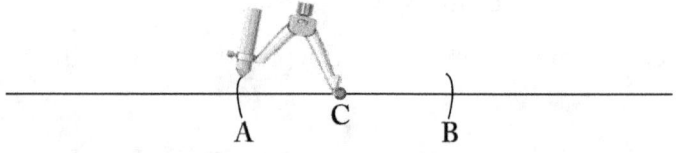

Step 3: With the same radius as step 2 and point A as the centre, draw an arc above and repeat the process for point B and call the point where both arcs meet point D.

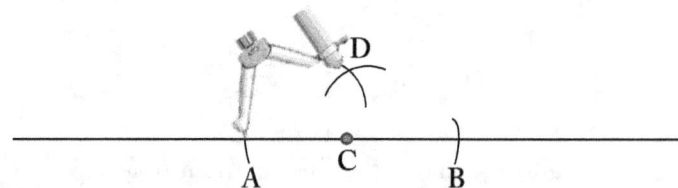

Step 4: Use your ruler to join point D to point C

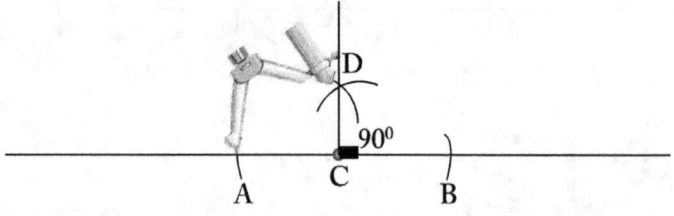

Step 5: The $\overline{CD}$ is perpendicular to $\overline{AB}$.

Mathematically, line CD being perpendicular to AB can be written as $\overline{CD} \perp \overline{AB}$

ALTERNATIVELY:

Step 1: Draw a straight line and mark point C above the line.

• C

_____

Step 2: With point C as the centre, draw an arc to cut the line the straight line at point A and B

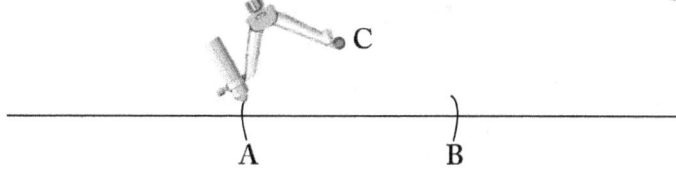

Step 3: With the same radius as step 2 and with point A as the centre, draw an arc below point A and repeat the same process for point B. Call the point where both arcs intersect or meet point D

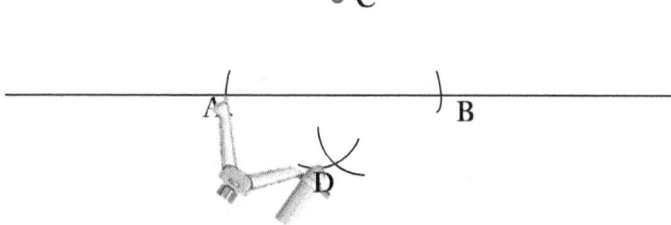

Step 4: Use your ruler to join point C to D, then the $\overline{CD}$ is perpendicular to $\overline{AB}$.

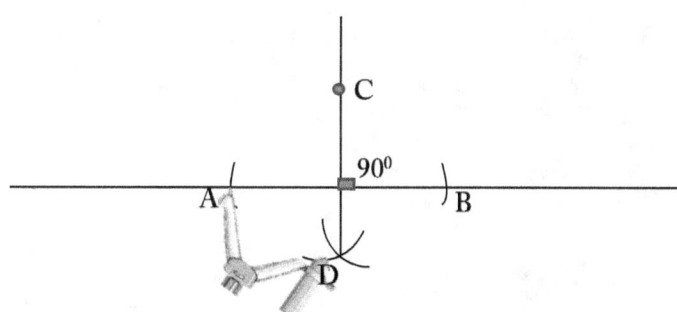

Mathematically, line CD being perpendicular to AB can be written as $\overline{CD} \perp \overline{AB}$

How To Draw Parallel Lines With Ruler And Set Squares

Step 1: Position an edge of the set square against a ruler and draw a line AB along one of the other edges.

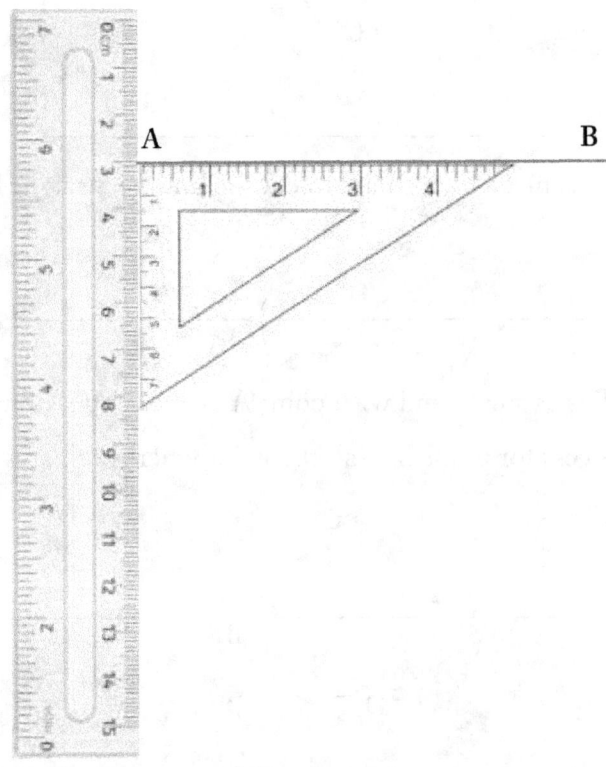

Step 2: Slide the set- square into a new position while keeping the ruler fixed exactly at the same position

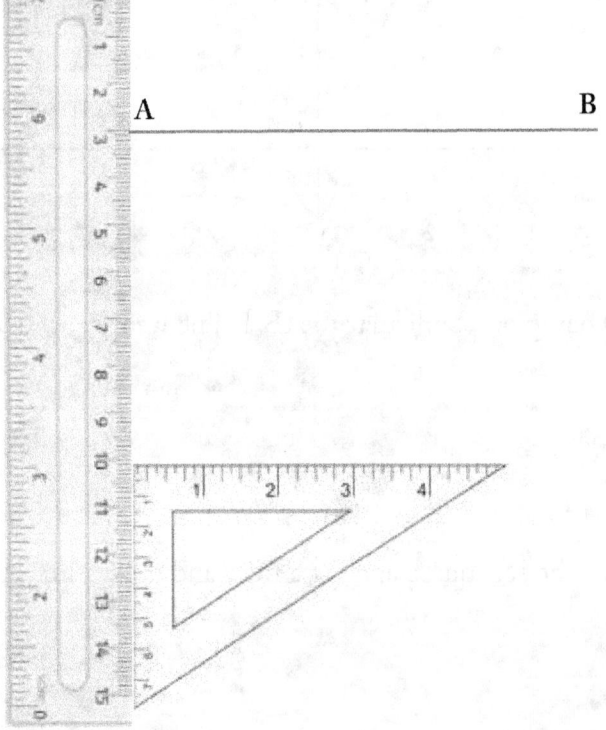

Step 3: Draw a line CD along the same edge just like the one in step 1

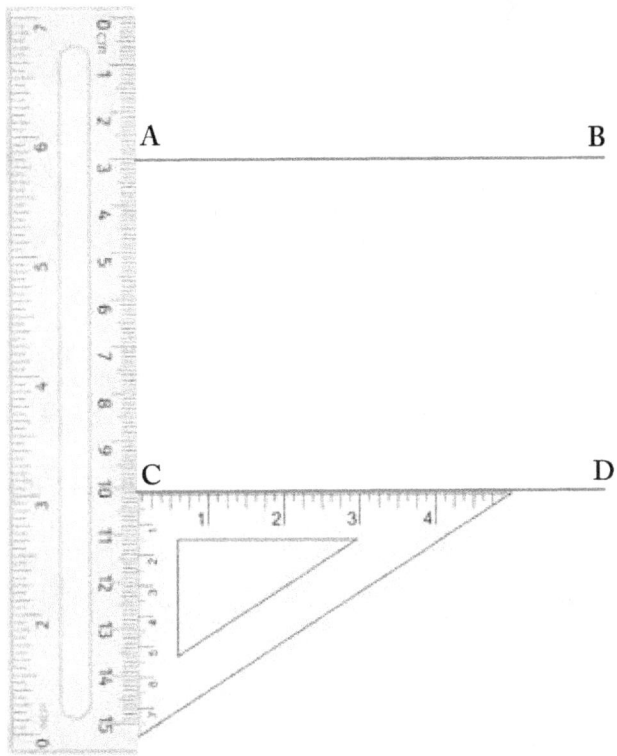

The line segment AB is parallel to CD mathematically, $\overline{AB} \parallel \overline{CD}$

How To Draw Perpendicular Lines With Ruler And Set Squares

Step 1: Draw line $\overline{AB}$, and mark point C on $\overline{AB}$

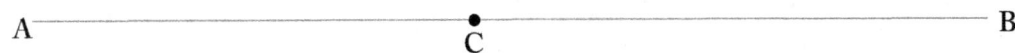

Step 2: Set an edge of the set square on $\overline{AB}$ so that the other edge is just in contact with point C.

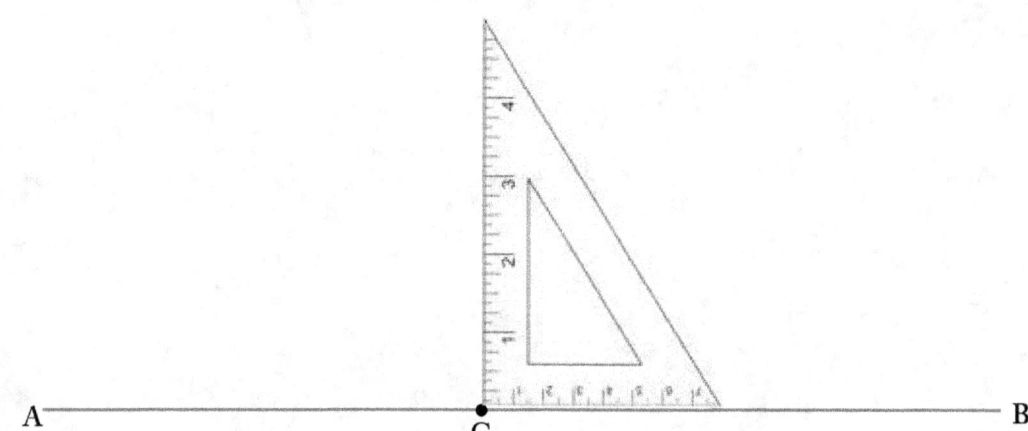

Step 3: Draw a line to pass through point C with the help of the set square.

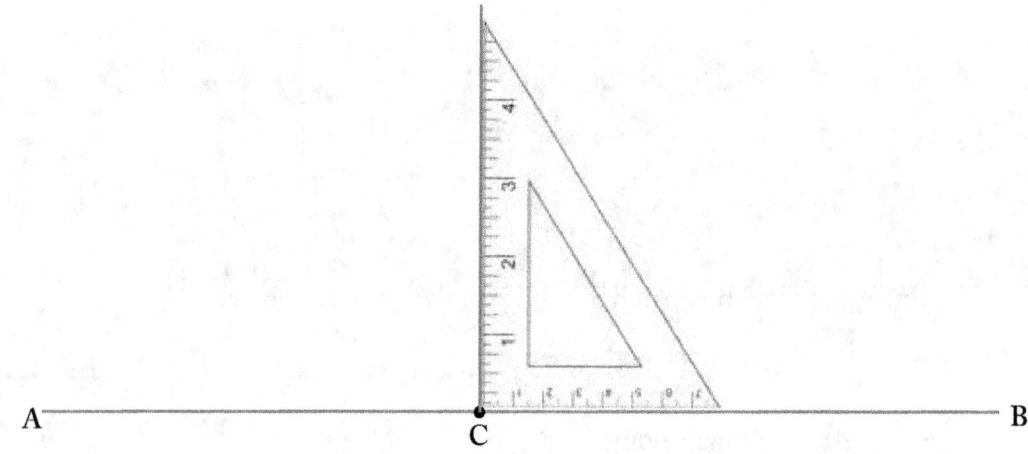

Exercise:

1. Draw two parallel line such that $|AB| = |CD| = 9cm$ and the distance between $|AB|$ and $|CD|$ is 6cm

2. Draw three parallel line such that $|AB| = |CD| = |EF| = 10cm$ and 4cm apart.

3. Draw a perpendicular line

4. Construct a parallel line

5. Draw $|XY| = 9cm$, bisect $\overline{XY}$.

CHAPTER FOUR

Angles

Definition of Angles

Angles are formed when two or more straight lines meet at a point. The common point where the straight lines meet is called *Vertex*. Angles are measured in degrees using a protractor.

Some angles can be constructed directly, some others are constructed by bisecting a previously constructed angles while some are gotten by adding up bisected angles. The angle that can be constructed directed are angles 60^0, 90^0 and 120^0. Like angle 30^0 can be gotten by bisecting (dividing into two equal part) angle 60^0 and angles 45^0 is from bisecting angle 90^0.

Examples of angles that can be constructed using a pair compass and ruler are 7.5^0, 15^0, 22.5^0, 30^0, 45^0, 60^0, 67.5^0, 75^0, 90^0, 105^0, 120^0, 135^0, 150^0, 165^0. Any angle apart from the angles listed above can only be done by using a protractor and a ruler to measure the angles, examples of such angles are 20^0, 40^0, 53^0, 70^0, 100^0 85^0 etc.

How To Bisect Angle Using Compass and Ruler

To bisect a given angle means to divide it into two equal parts. An angle bisector is a straight line that divides the angle into two equal parts while a vertex is the point where both straight lines making up the angles meet.

How To Bisect Angle ∠ACB With Vertex C

Procedures:

Step 1: Place the sharp point of your compass at point C (the vertex C), and with any convenient radius of your choice, draw an arc to cut line CA at point D and line CB at point E.

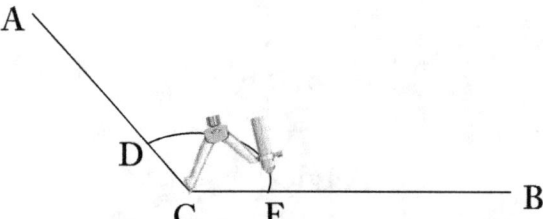

Step 2: Place the sharp point of the compass at point D, with the same radius draw an arc above the arc DE

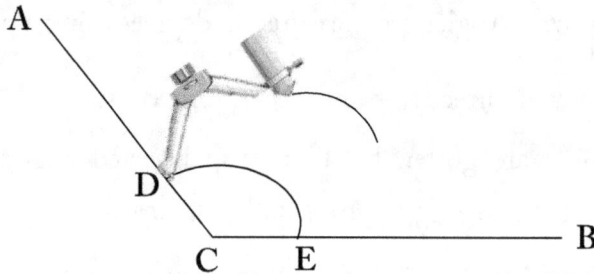

Step 3: With the same radius as the previous steps, place the sharp point of your compass at point E, draw

another arc to meet the first arc in step 2, and call the point where both arcs meet point F.

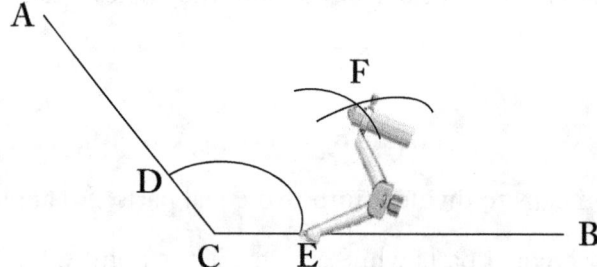

Step 4: Use your ruler to join the points CF together. The line CF is the bisector of ∠ACB

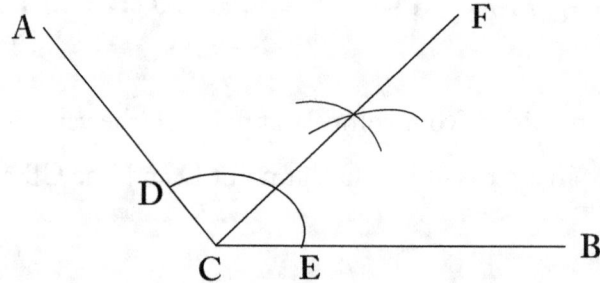

∠FCB = ∠ACF

Construction of Angles

In this section, we will learn how to construct angles 60^0, 30^0, 90^0, 45^0, 120^0, 150^0, 75^0. Subsequently we will learn how to bisect the afore-mentioned angles.

How to Construct Angle 60^0

How to construct angle 60^0

Step 1: Draw a straight-line AB of any length

Step 2: Place the sharp point of your compass at point A, and with any convenient radius draw an arc to cut AB at point C

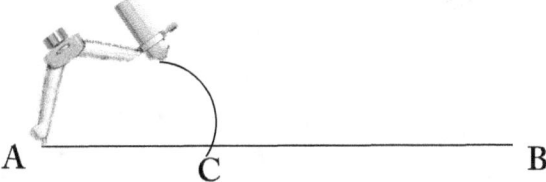

Step 3: With the same radius as step 2, place the sharp point of your compass on point C, draw an arc to cut the first arc at point D

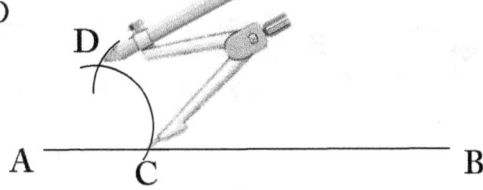

Step 4: Use your ruler to join point A to D and extend it to point E. Therefore, $\angle EAB = 60^0$

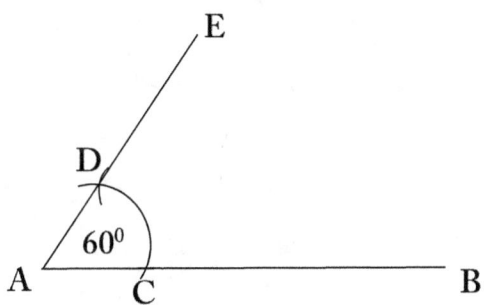

How to construct angle 30^0

Step 1: first construct angle 60^0 as stated above

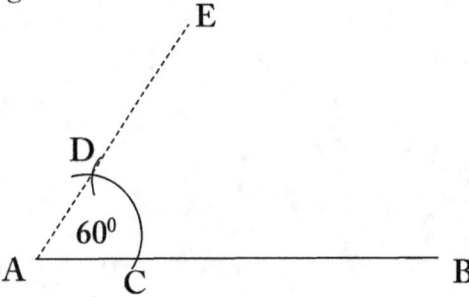

Step 2: Bisect angle 60^0 to get angle 30^0 (with the same radius that was used to construct 60^0, place the sharp point of the compass at point C, draw an arc. Repeat the same process for point D and where both arcs cut each other call it point F)

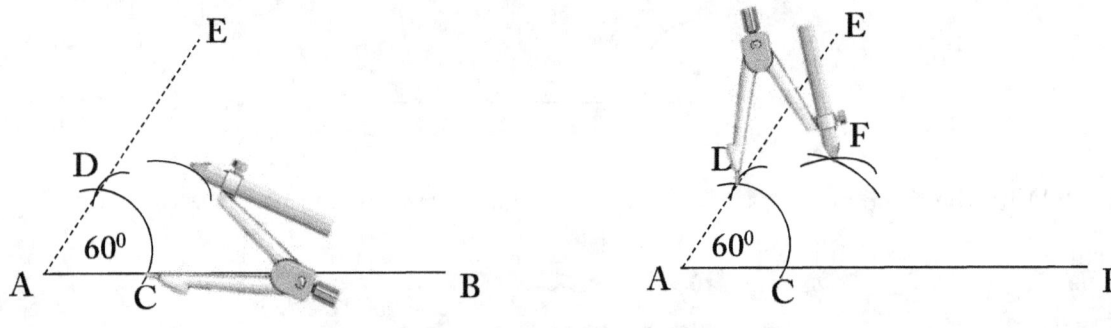

Step 3: Use your ruler to draw a straight line to join point F and A together. Therefore, $\angle FAB = 30^0$

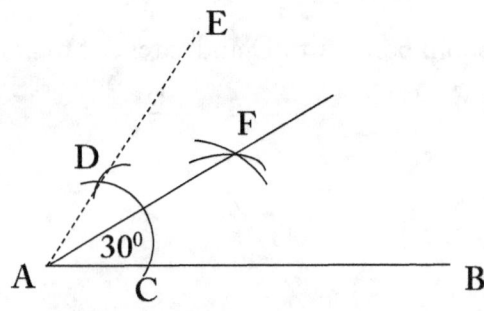

How to construct angle 15⁰

Step 1: First construct angle 60⁰ as stated above

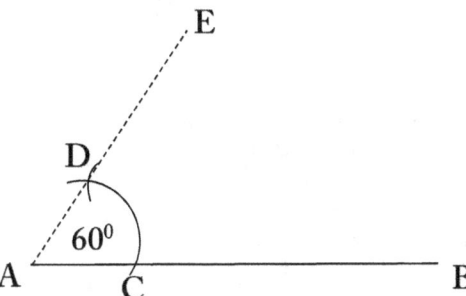

Step 2: Bisect angle 60⁰ to get angle 30⁰ as shown above

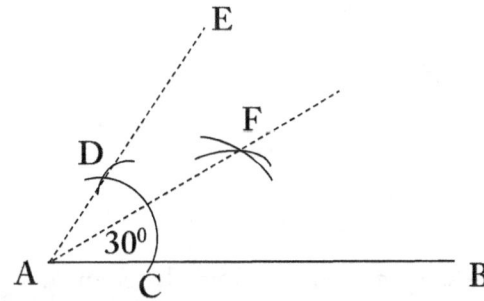

Step 3: Bisect angle 30⁰ to get angle 15⁰

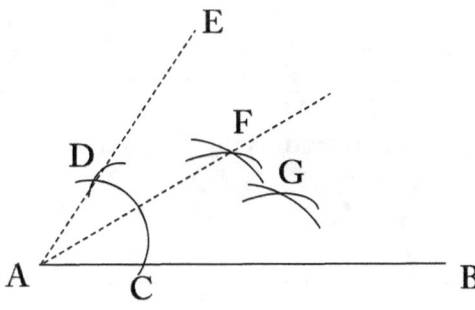

Step 4: Use your ruler to join point A to G and extend it. Therefore, $\angle GAB = 15^0$

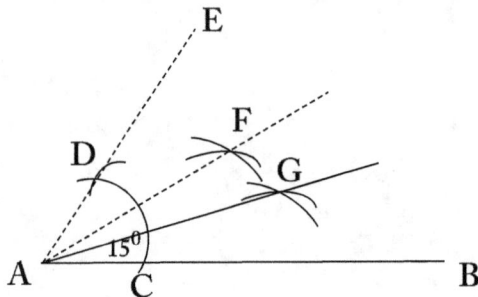

How to Construct Angle 7.5⁰

How to construct angle 7.5⁰

Step 1: First construct angle 60^0 as stated above

Step 2: Bisect angle 60^0 to get angle 30^0

Step 3: Bisect angle 30^0 to get angle 15^0

Step 4: bisect angle 15⁰ to get angle 7.5⁰

How to Construct Angle 90⁰

The construction of angle 90^0 is the same as constructing a bisector of a given straight line.

How to construct angle 90⁰

Step 1: Draw a straight line and mark point C at the centre.

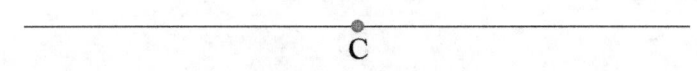

Step 2: With C as the centre, draw an arc of any radius on both sides of point C and call it A and B.

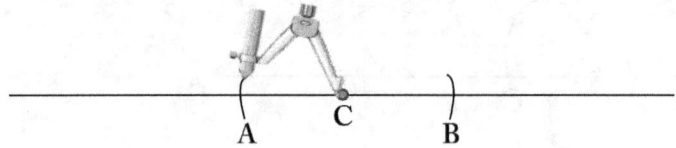

Step 3: With the same radius as step 2 and point A as the centre, draw an arc above and repeat the process for point B, and call the point of intersection of both arcs D.

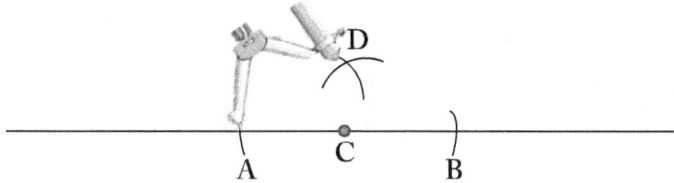

Step 4: Use your ruler to join point D to point C

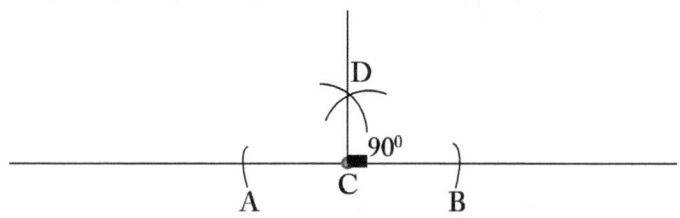

How to Construct Angle 45⁰

How to construct angle 45⁰

Step 1: First construct angle **90⁰** as stated above

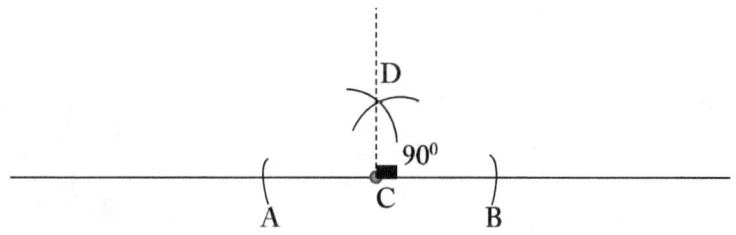

Step 2: Bisect angle **90⁰** to get angle **45⁰**

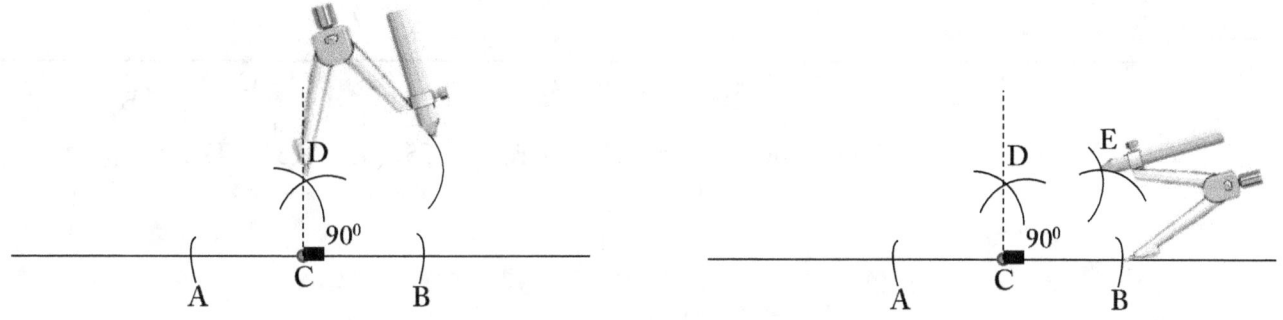

Step 3: Use your ruler to join point C to E and extend it. Therefore, $\angle ECB = 45^{0}$

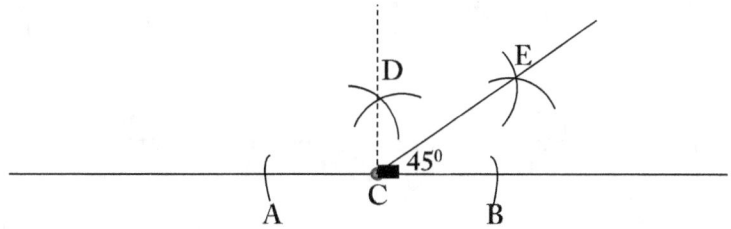

How to construct angle 22.5⁰

Step 1: First construct angle **90⁰** as stated above

Step 2: Bisect angle **90⁰** to get angle **45⁰**

Step 3: Bisect angle **45⁰** to get angle **22.5⁰**

How to Construct Angle 75⁰

How to construct angle 75⁰

Step 1: First construct angle **90⁰** and **60⁰** stated above

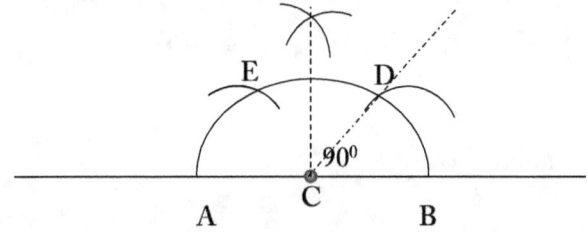

Step 2: Bisect the space between angle **90⁰** and **60⁰** (i.e. bisect the space points F and D) to get your angle **75⁰**

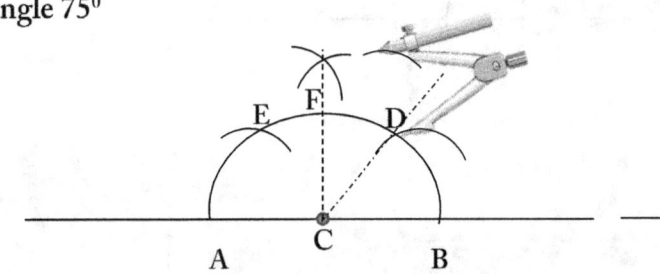

Step 3: Add the space to that of angle **60⁰** to make your angle **75⁰** (**60⁰ + 15⁰ = 75⁰**)

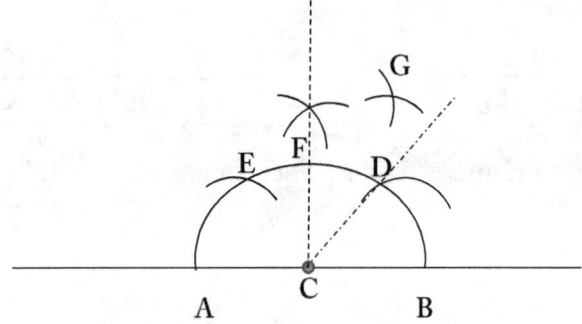

Step 4: Use your ruler to join point C to G and extend it. Therefore, ∠*GCB* = **45⁰**

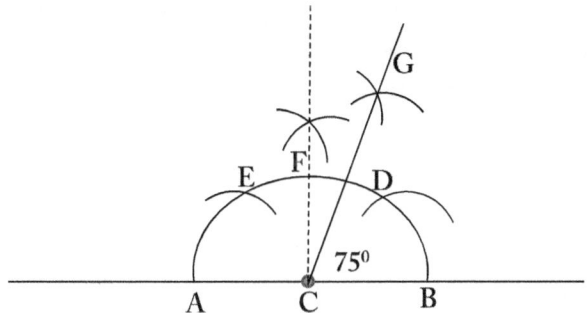

How to Construct Angle 120⁰

How to construct angle 120⁰

Step 1: Draw a straight-line AB of any length

A   B

Step 2: Place the sharp point of your compass at point A, and with any convenient radius draw an arc to cut AB at point C.

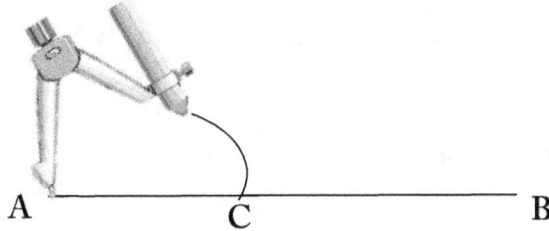

Step 3: With the same radius as step 2, place the sharp point of your compass on point C, draw an arc to cut the first arc at point D.

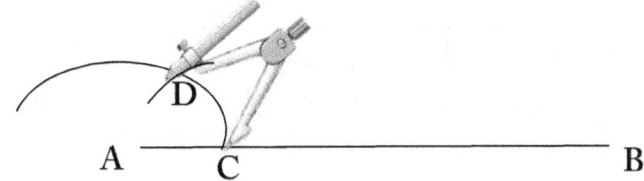

Step 4: With the same radius above, place the sharp point of your compass on point D, draw an arc to cut the previous arc at point E.

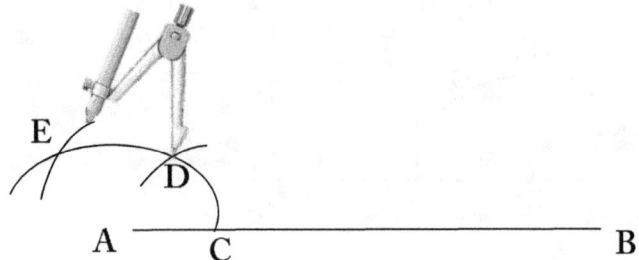

Step 5: Use your ruler to join point A to E and extend it to point F. Therefore, $\angle FAB = 120^0$

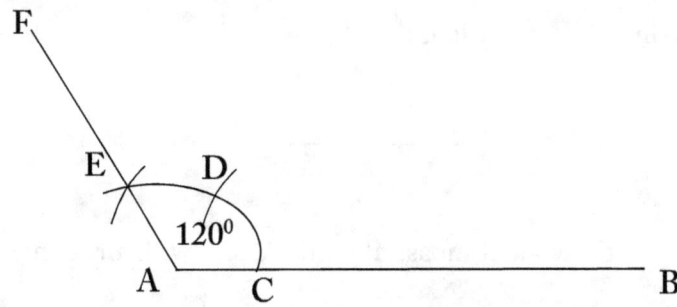

How to Construct Angle 105^0

How to construct angle 105^0

Step 1: First construct angle 90^0 and 120^0 stated above

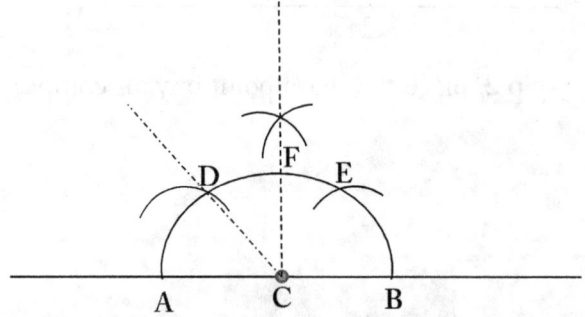

Step 2: Bisect the space between angle 90^0 and 120^0 (i.e. the space between point D and F) to get your

angle 105^0

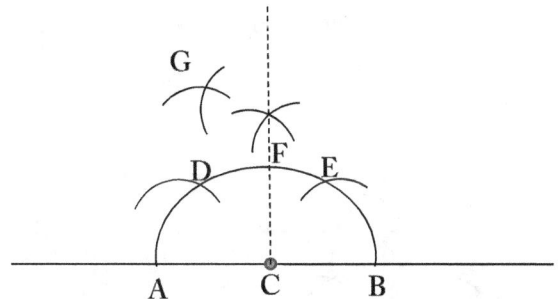

Step 3: Add the space to that of angle 90⁰ to make your angle 105⁰ (90⁰ + 15⁰ = 105⁰)

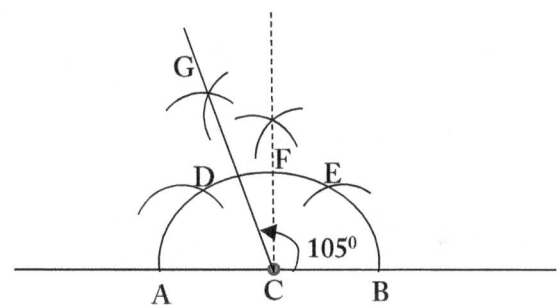

How to Construct Angle 135⁰

How to construct angle 135⁰

Step 1: First construct angle 90⁰

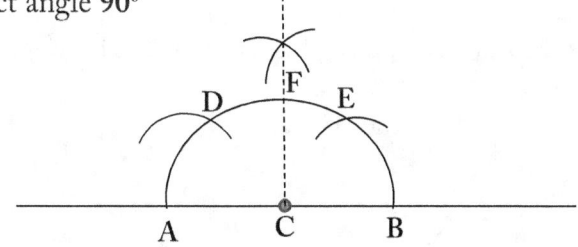

Step 2: Bisect the second angle 90⁰ to get angle 45⁰ **on the left.**

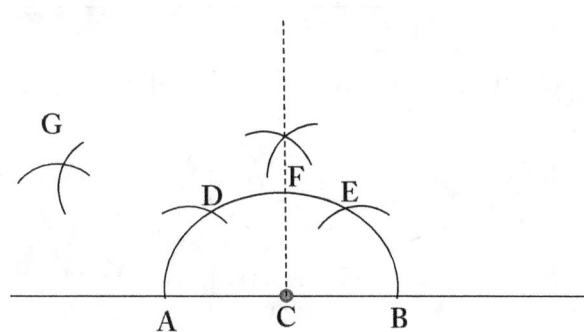

Step 3: Add the angle 45⁰ to the angle 90⁰ to make your angle 135⁰ (90⁰ + 45⁰ = 135⁰)

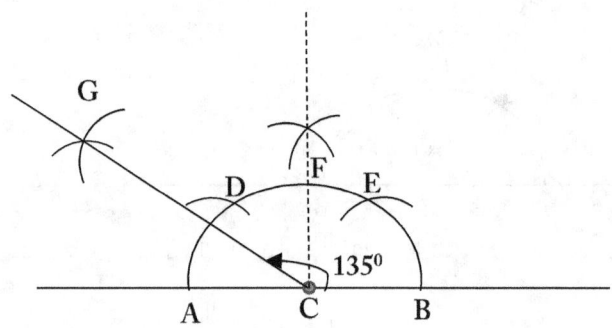

How to Construct Angle 150⁰

How to construct angle 150⁰

Step 1: First construct angle 120⁰ as shown above

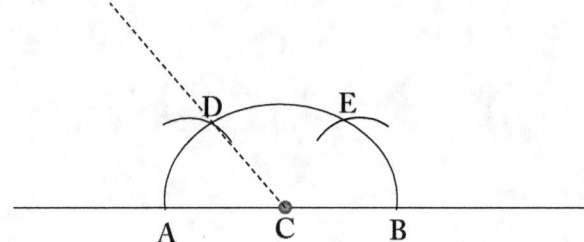

Step 2: With the same radius as the one used above, bisect the space between point A and D, call the

point where both arcs intersect each other F.

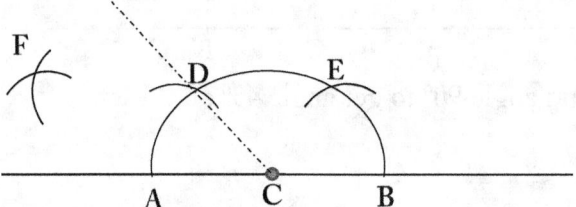

Step 3: Use your ruler to join point F to C and extend it. Therefore, ∠*FCB* = **150⁰**

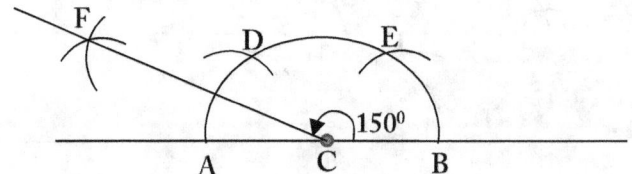

How to construct of angle 165^0

Step 1: First construct angle 150^0 as shown above

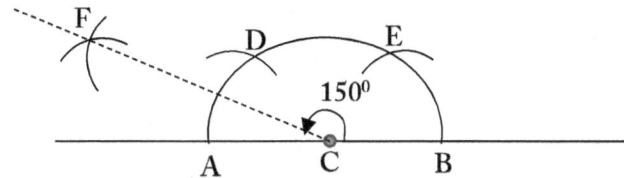

Step 2: With the same radius as the one used above, bisect the remaining angle 30^0

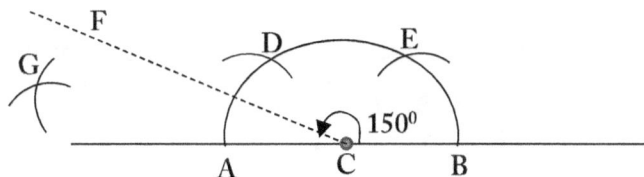

Step 3: Use your ruler to join point G to C and extend it. Therefore, $\angle GCB = 165^0$

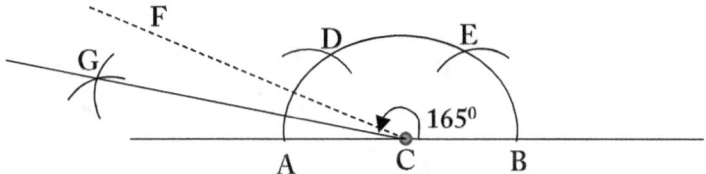

Exercise

Using a ruler and a pair of compasses construct the following:

1. Construct 60^0

----

2. Construct 30^0

----

3. Construct 15^0

4. Construct 7.5^0

- -

5. Construct 90^0

- -

6. Construct 45^0

7. Construct 22.5^0

----

8. Construct 75^0

9. Construct 120^0

10. Construct 135^0

11. Construct 150^0

12. Construct 165^0

Measure and record the following angles using your protractor

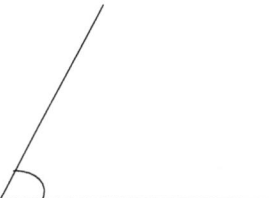

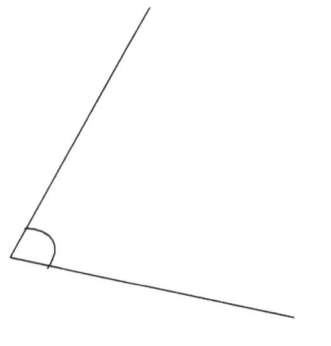

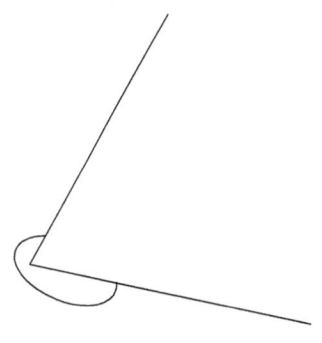

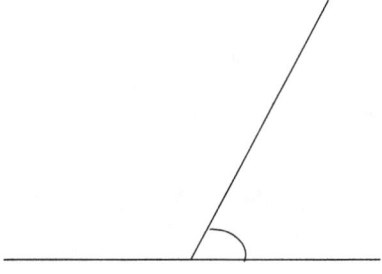

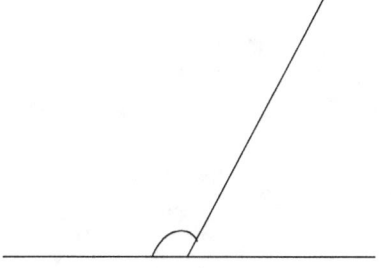

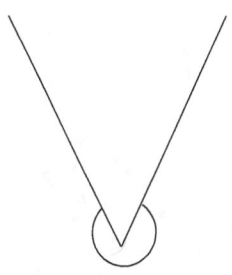

CHAPTER FIVE

Construction of Triangles

Meaning of Triangles

The word "tri" means three and therefore a figure with three angles is called a triangle. A triangle is a closed plane figure with three straight edges, three angles and three vertices.

Types of Triangles

Triangles are classified into two broad categories based on their *"internal angles"* and *"sides"*.

By side: equilateral triangle, isosceles triangle, and scalene triangle. By internal angles: acute angled triangle, right angled triangle and obtuse or oblique angled triangle.

By Side	By Angle
Equilateral Triangle Has three equal sides	**Acute angled triangle** Has all its angles less than 90^0
Isosceles triangle Has two equal sides	**Right angled triangle** Has one of its angles equals 90^0
Scalene Triangle Has no equal sides	**Obtuse /Oblique angled triangle** Has one of its angles greater than 90^0

How To Construct Equilateral Triangle

Equilateral triangle is a triangle with three equal side and three equal angles (each angle is 60^0).

Construct a ∆ABC with $|AB| = |AC| = |BC| = 7cm$

Procedure:

Step 1: Draw a line $|AB| = 7cm$

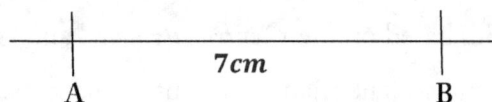

7cm

A   B

Step 2: With A as the centre and radius $7cm$ draw an arc above $|AB|$

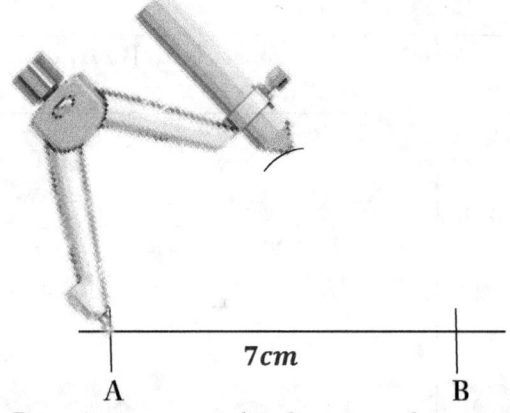

7cm

A   B

Step 3: With B as the centre and radius $7cm$ draw an arc above to cut the previous arc and call the

intersection C.

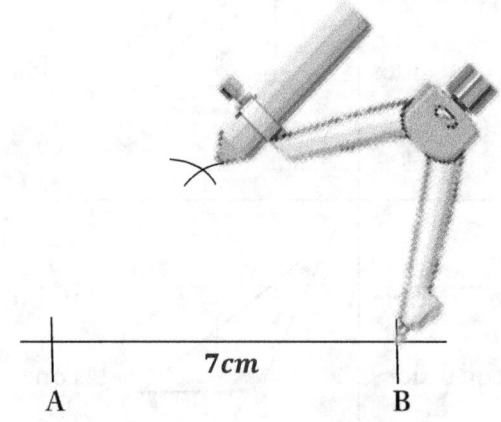

7cm

A   B

Step 4: Use your ruler to join AC and BC to get the required triangle ABC.

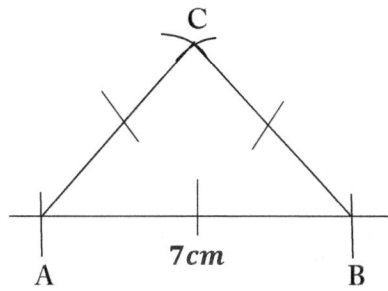

How To Construct Isosceles Triangle

An isosceles triangle has two sides that are equal and their base angles are equal.

Example

Construct a triangle ABC with $|AB| = |AC| = 7cm, |BC| = 5cm$.

Procedures:

Step 1: Draw a line such that $|BC| = 5cm$

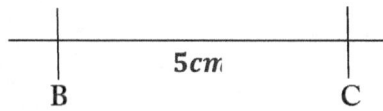

Step 2: With B as the centre and radius $7cm$ draw an arc above $|BC|$

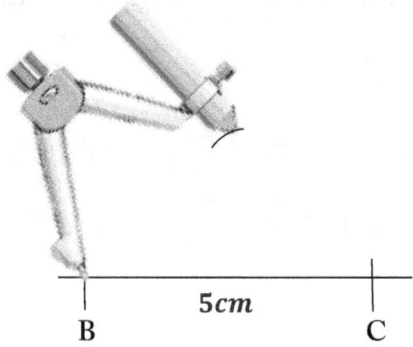

Step 3: With C as the centre and radius $7cm$, draw another arc above to cut the previous arc and call the intersection A

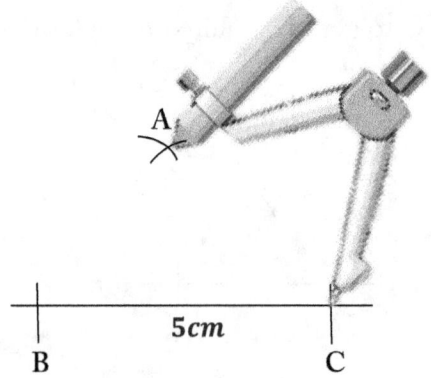

5cm

B   C

Step 4: Use your ruler to join AB and AC to get the required triangle ABC.

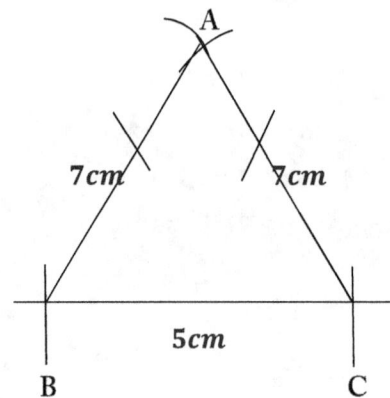

7cm   7cm

5cm

B   C

How To Construct Scalene Triangle

A scalene triangle is a triangle with three unequal sides and three unequal angles.

Example

Construct a triangle ABC with $|AB| = 5cm, |BC| = 6cm, and |AC| = 7cm.$

Procedures:

Step 1: draw a line such that $|AB| = 5cm$

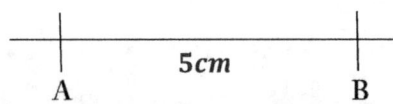

5cm

A   B

Step 2: with A as the centre and radius $7cm$ draw an arc above $|AB|$

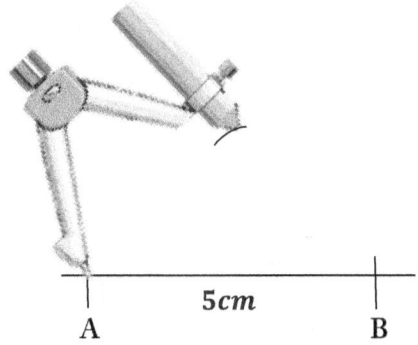

Step 3: With B as the centre and radius 6cm, draw another arc above to cut the previous arc and call the intersection C.

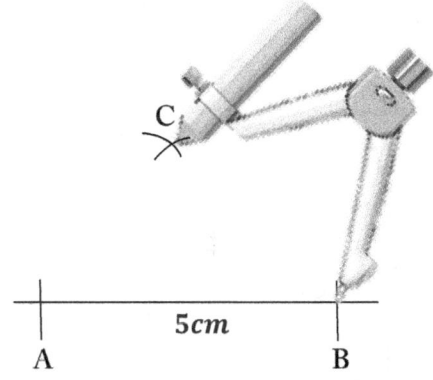

Step 4: use your ruler to join AC and BC to get the required triangle ABC.

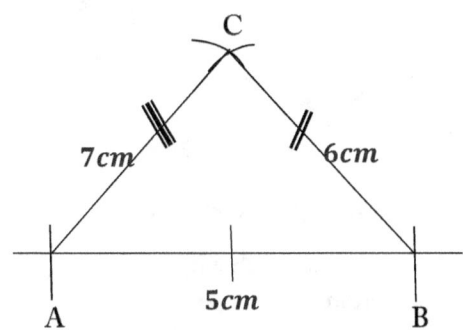

How To Construct A Triangle Given Two Angles And One Side

Example:

Construct $\triangle ABC$ such that $|AC| = 7.5cm, < CAB = 60^0$ *and* $\angle ACB = 45^0$ *Measure* $|AB|$

Procedures

Step 1: Draw a line $|AC| = 7.5cm$

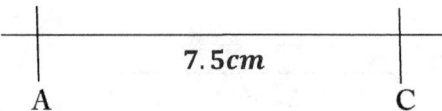

7.5cm

A   C

Step 2: With A as the centre, construct angle 60^0 at point A ($\angle BAC = 60^0$ i.e. angle 60^0 is at point A)

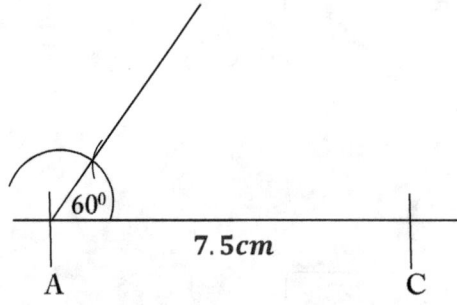

Step 3: With C as the centre, construct angle 45^0 ($\angle ACB = 45^0$) and mark the point where the arms for angle 60^0 and 45^0 meet as B to get the required triangle.

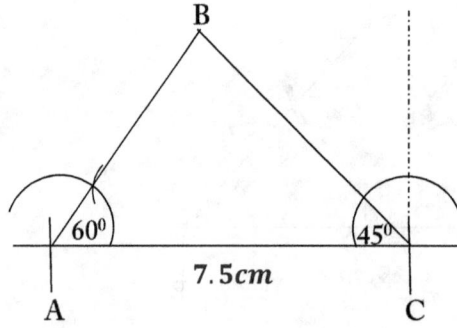

How To Construct A Triangle With One Angle And Two Sides

Example:

Construct $\triangle ABC$ such that $|AB| = 7cm, |BC| = 10cm, \angle ABC = 60^0$

Procedure:

Step 1: Draw a line $|AB| = 7cm$

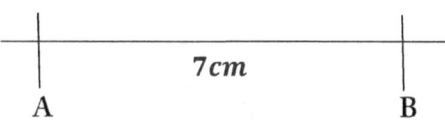

Step 2: With B as the centre, construct angle 60^0

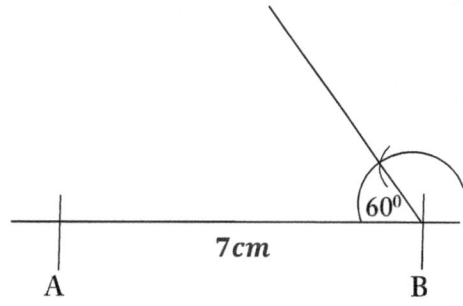

Step 3: With B as the centre and radius $10cm$, draw an arc to cut the arm angle 60^0 at point C.

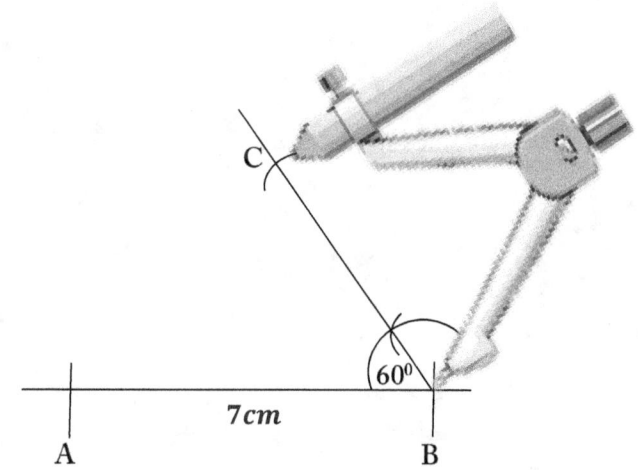

Step 4: Use ruler to join point A to C to get the required triangle.

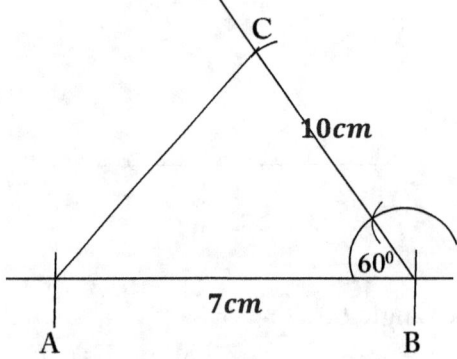

Exercise

1. Draw $\triangle ABC$ when $|AB| = 5cm$, $\angle A = 40^0$ and $< B = 50^0$

----

2. Construct a $\triangle ABC$ in which $|AB| = |AC| = 7.2cm$, and $|AB| = 9cm$.

3. Construct a ΔPQR in which $|PQ| = 5.8cm$ $|QR| = 6.5cm$, and $= 4.5cm$.

----

4. Using a ruler and a pair of compasses only construct ΔXYZ with sides $|XY| = 8cm$ $\angle XYZ = 60^0$ and $\angle YXZ = 45^0$. Measure $|YZ|$ and $|XZ|$

5. Using a ruler and a pair of compasses only, construct $\triangle ABC$ with side $|AB| = |BC| = |AC| = 8cm$

----

6. Using a ruler and a pair of compasses only, construct $\triangle ABC$ with side $|AB| = 7.5cm$ $|BC| = 6cm$ $|AC| = 10cm$

CHAPTER SIX

Constructing Quadrilateral

Meaning of Quadrilateral

The word quadrilateral originated from two Latin words *quadri* which means *"four"* and *latus* meaning *"side."* A quadrilateral is a closed plane figure with four straight sides, four angles and four vertices. The sum of all the interior angles of a quadrilateral is always 360^0. Quadrilateral can be squares, rectangles, kites, rhombuses, parallelograms and trapeziums.

SQUARE

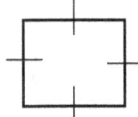

Properties:

    i.   All the four sides are equal.

    ii.   Opposite sides are equal and parallel.

    iii.   All the four angles are equal and measure 90^0 (right angle)

    iv.   The diagonals are equal and they bisect each other at the centre to form right angle.

RECTANGLE

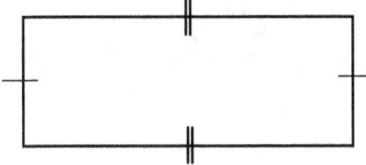

Properties:

    i.   Opposite sides are equal and parallel.

    ii.   All the four angles are equal and measure 90^0

    iii.   The diagonals are equal and they bisect each other

    iv.   The diagonals of a rectangle bisect each other at different angles- One acute and the other obtuse.

v.   A rectangle whose diagonals bisect each other at the centre to form right angle is called a square.

RHOMBUS

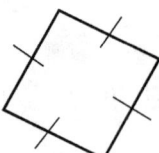

Properties:

i.   All the four sides are equal.

ii.   Opposite sides are equal and parallel.

iii.   The diagonals bisect each other at right angle.

iv.   The diagonal bisects the angle of a rhombus.

v.   Opposite angles are congruent (equal).

vi.   The sum of any two adjacent or consecutive angle is 180^0

vii.   The diagonals are not equal, but they bisect each other at the centre

PARALLELOGRAM

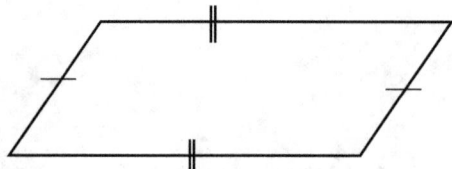

Properties:

i.   Opposite sides are parallel and congruent.

ii.   The diagonals bisect each other.

iii.   The opposite angles are congruent (equal)

iv.   Each diagonal of a parallelogram separates it into two congruent triangles

v.   Consecutive angles are supplementary

TRAPEZIUM

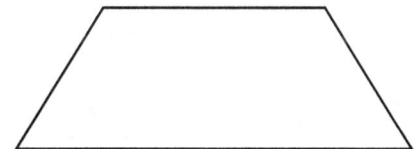

Properties:

    i.   It has a pair of non-parallel line

    ii.   It has a pair parallel side.

    iii.   The length of both diagonals is equal

    iv.   Two angles on the same side is always supplementary i.e. sum up to 180^0

    v.   The diagonals bisect each other.

How To Construct A Square

How to construct a square given the length of one side

Example

Construct a square ABCD such that $|AB| = 5cm$

Procedures:

Step 1: Draw a line $|AB| = 5cm$ and extend the line through B

A   **5cm**   B

Step 2: Construct a perpendicular at point B

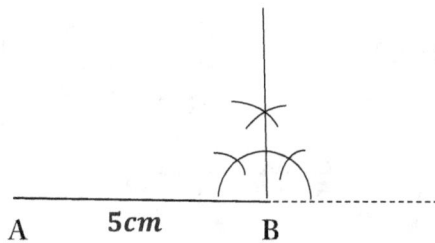

A   **5cm**   B

Step 3: With B as the centre and radius 5cm, draw an arc to cut the arm of the perpendicular line and call it C.

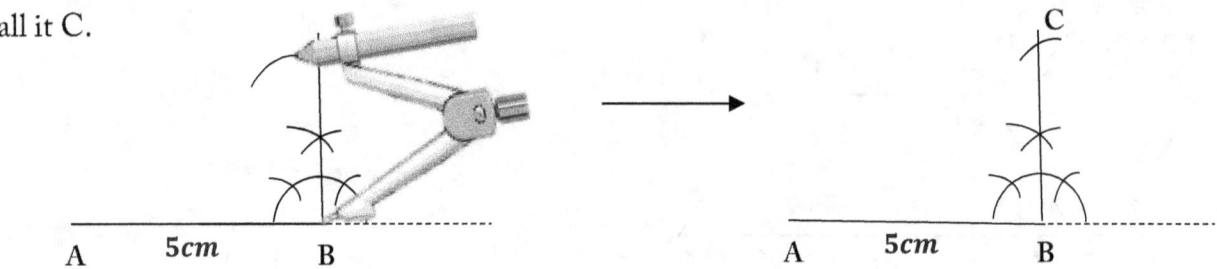

Step 4: With C as the centre and radius 5cm, draw an arc

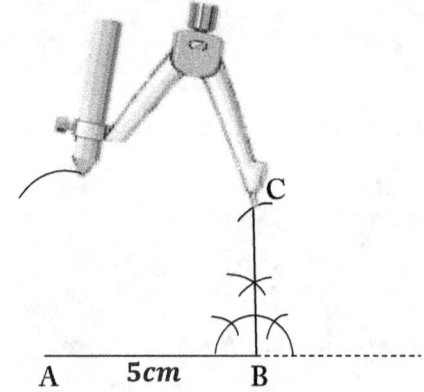

Step 5: With A as the centre and radius 5cm draw an arc to cut the previous arc and call it D

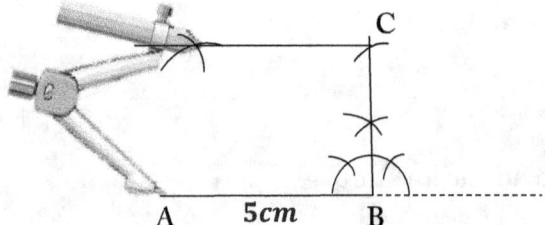

Step 6: Use your ruler to connect all the four points together to get the required square.

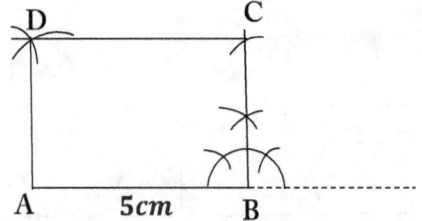

How to construct a square inscribed in a circle

Step 1: Use your compass to draw a circle with centre O

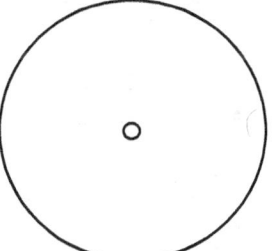

Step 2: Use a ruler, draw a diameter of the circle and call the endpoint A and B

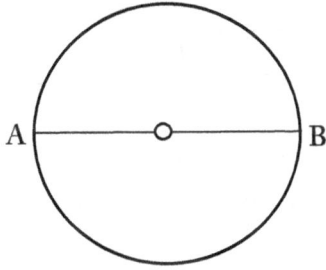

Step 3: Construct a perpendicular bisector of the diameter AB and call points where it intersects the circle C and D (i.e. with A and B as centre draw an arc above and below the circle)

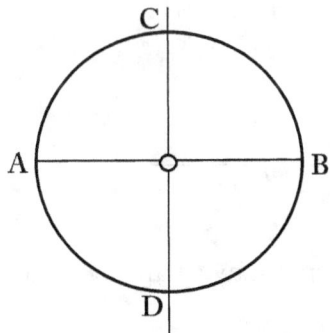

Step 5: Joint the points A, B, C and D to form a square.

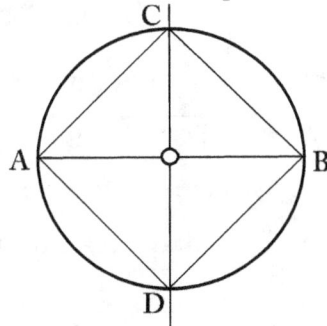

How To Construct A Rectangle

How to construct a rectangle when two adjacent sides are given

Example:

Construct a rectangle ABCD such that $|AB| = 5cm$ and $|BC| = 3cm$

Procedures:

Step 1: Draw a line such that $|AB| = 5cm$

A   **5cm**   B

Step 2: With A as the centre, construct angle 90^0, (all the angles in the rectangle is right angle each)

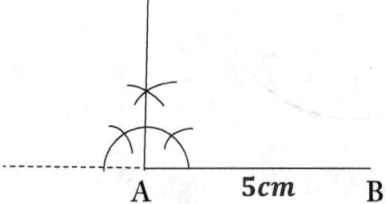

A   **5cm**   B

Step 3: With radius $3cm$ and A as the centre draw an arc to cut the perpendicular bisector of AB at point D.

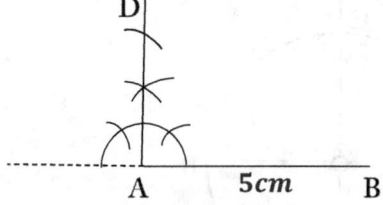

A   **5cm**   B

Step 4: With B as the centre and radius 3cm, draw an arc above B

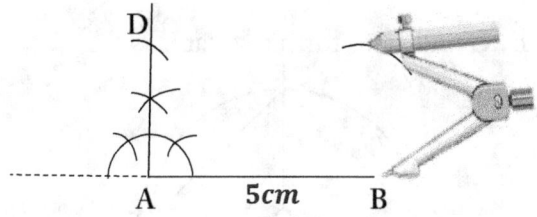

A   **5cm**   B

Step 5: With D as the centre and radius of $5cm$, draw an arc to cut the previous arc above B and call the points where both arcs intersect C.

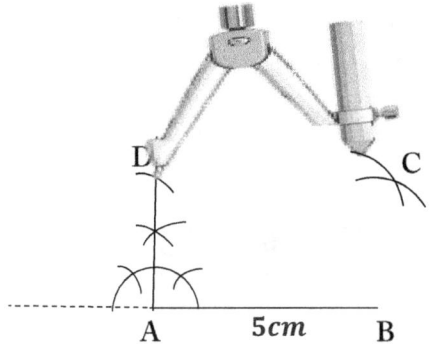

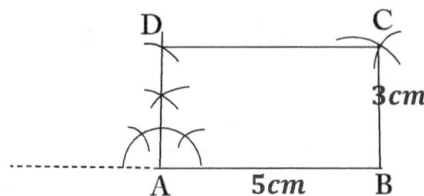

Step 6: Use ruler to join all the points to form the required rectangle.

How To Construct A Rhombus

How to construct a rhombus given one side and one diagonal.

Example:

Construct a rhombus ABCD such that $|AB| = 6cm$ and diagonal $|AC| = 9cm$

Procedures:

Step 1: Draw a line such that $|AB| = 6cm$

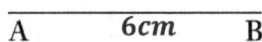

Step 2: With B as the centre and radius $6cm$, draw an arc above point B

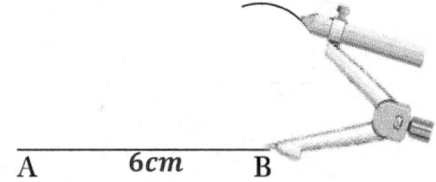

Step 3: With A as the centre and radius 9*cm*, draw another arc to intersect the previous arc and call it C.

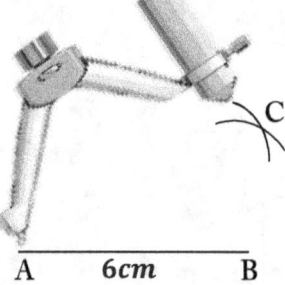

Step 4: With A as the centre and radius 6*cm*, draw an arc above

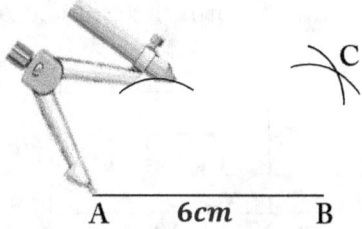

Step 5: With C as the centre and radius 6*cm*, draw another arc to intersect the previous arc and call it D.

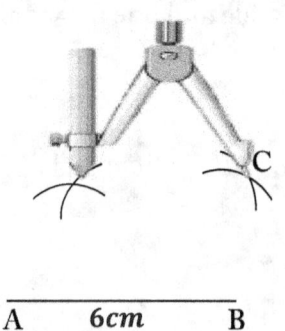

Step 6: Use ruler to join all the points together to form the rhombus

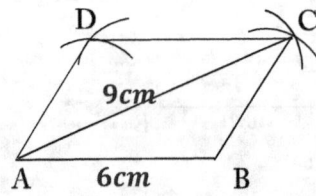

How to construct a rhombus giving one side and one angle.

Example:

Construct a rhombus ABCD such that $|AB| = 6cm$ and $\angle A = 60^0$

Procedures:

Step 1: Draw a line such that $|AB| = 6cm$

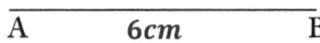

A   6*cm*   B

Step 2: With A as the centre, construct angle 60^0

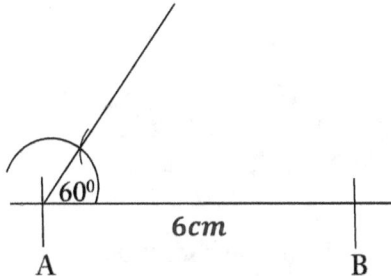

Step 3: With A as the centre and radius 6*cm* draw an arc to cut the arm angle 60^0 and call it D

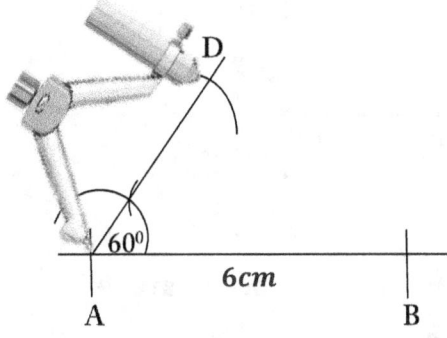

Step 4: With D as the centre and radius 6*cm* draw an arc.

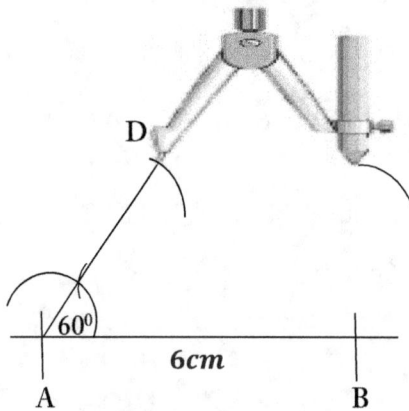

Step 5: With B as the centre radius 6*cm*, draw an arc to cut the previous arc and call it C

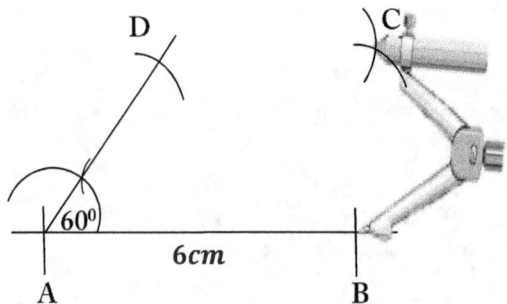

Step 6: Use ruler to join all the points together to form the rhombus.

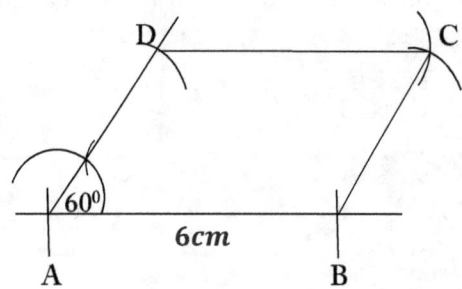

How To Construct A Parallelogram

How to Construct a parallelogram when two consecutive sides and a diagonal are given.

Example

Construct a parallelogram ABCD in which $|AB| = 6cm$, $|BC| = 5cm$ and diagonal $|AC| = 7cm$

Procedures:

Step 1: Draw a line $|AB| = 6cm$

A   6cm   B

Step 2: With A as the centre and radius 7cm, draw an arc above.

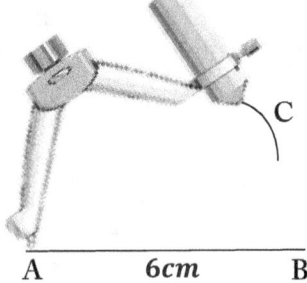

A   6cm   B

Step 3: With B as the centre and radius 5cm, draw another arc to cut the previous arc at C and join BC and AC.

Step 4: With A as the centre and radius 5cm, draw an arc

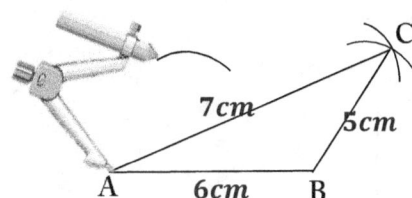

Step 5: With C as the centre and radius 6cm draw another arc cutting the previous arc and call it D

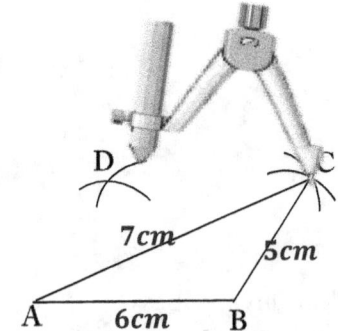

Step 6: Use ruler to join DA and DC to form the required parallelogram.

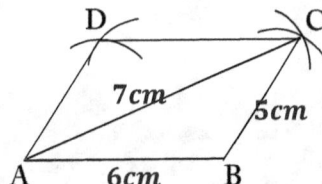

How to Construct a parallelogram when two consecutive sides and the angle included are given.

Example:

Construct a parallelogram ABCD with sides $|AB| = 5cm$, $|AD| = 7cm$ and $\angle A = 60^0$

Step 1: Draw a line $|AB| = 5cm$

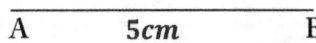

A   5cm   B

Step 2: Construct angle 60^0 at point A

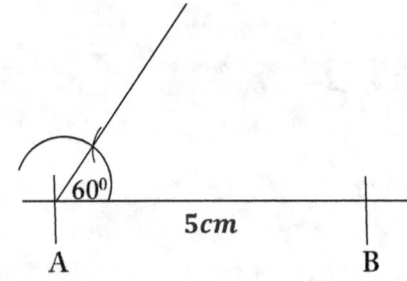

Step 3: With A as the centre and radius 7cm construct an arc to cut the arm of the angle and call it D.

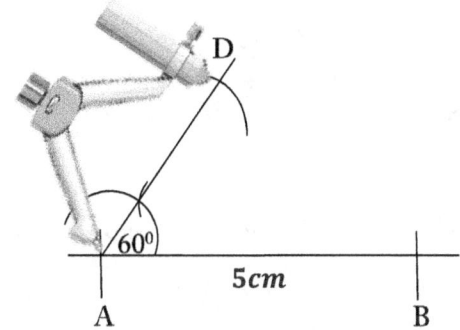

Step 4: With B as the centre and radius 7cm, draw an arc above.

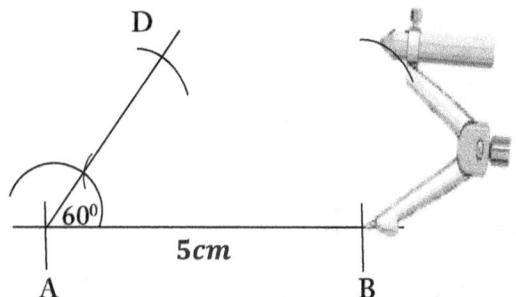

Step 5: With D as centre and radius 5cm draw an arc to cut the previous arc and call it C

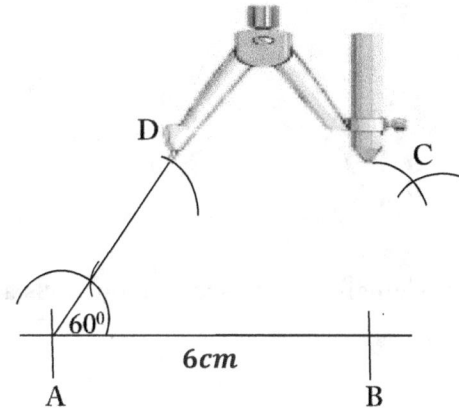

Step 6: Use ruler to join all the points together to form the required parallelogram.

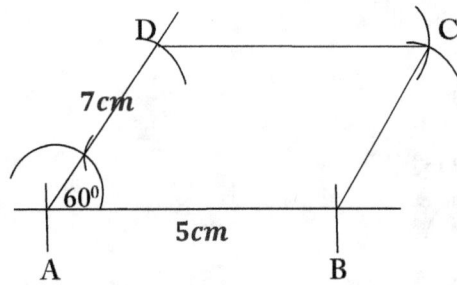

How to construct a parallelogram when one side and two diagonals are given.

Example:

Construct a parallelogram ABCD in which $|AB| = 7cm$, and diagonals $|BD| = 8cm$, $|AC| = 10cm$

Procedures:

Step 1: Draw a line $|AB| = 7cm$

A ———— 7cm ———— B

Step 2: With A as the centre and radius 5cm, draw an arc

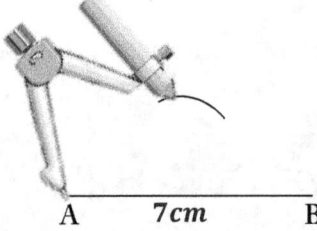

A   7cm   B

Step 3: With B as the centre and radius 4cm, draw an arc to intersect the previous arc and call it O, and join AO and BO to form a triangle.

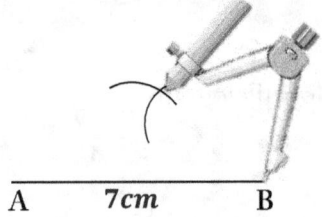

A   7cm   B

Step 4: With O as the centre and radius 5cm draw an arc and call it C.

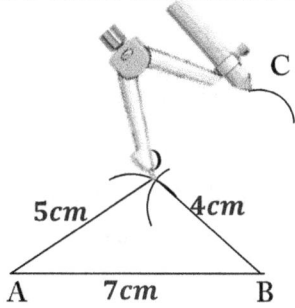

Step 5: With O as the centre and radius 4cm draw an arc

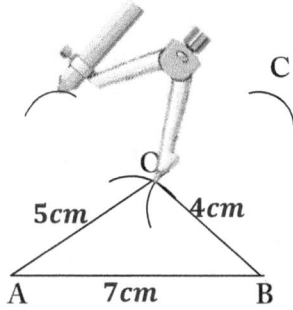

Step 6: With C as the centre and radius 7cm, draw an arc to cut the previous arc and call it D

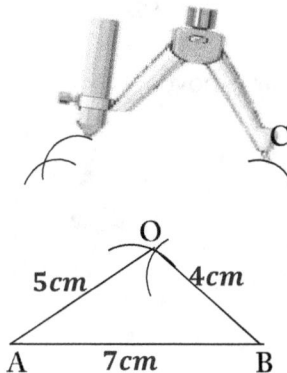

Step 7: Use ruler to join all the points together to form the required parallelogram.

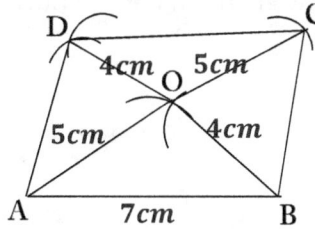

How To Construct A Trapezium

How to construct a trapezium when all the four sides are given

Example:

Construct a trapezium ABCD in which $|AB| = 7cm$, $|BC| = |CD| = 4cm$, and $|DA| = 5cm$. $|AB|$ *paralell* $|CD|$

Procedures:

Step 1: Draw a line $|AB| = 7cm$

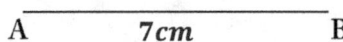

Step 2: With A as the centre and radius 4cm, draw an arc to cut line AB at point E

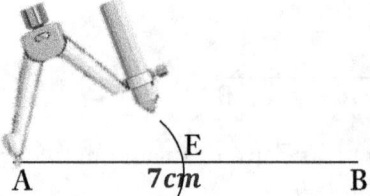

Step 3: With E as the centre and radius 5cm draw an arc above.

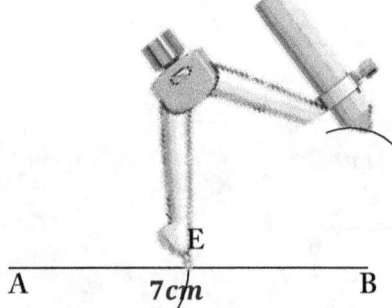

Step 4: With B as centre and radius 4cm, draw an arc to cut the previous arc at C.

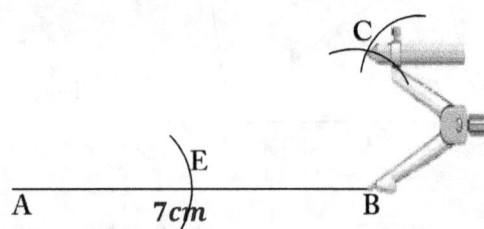

Step 5: With C as the centre and radius 4cm, draw an arc

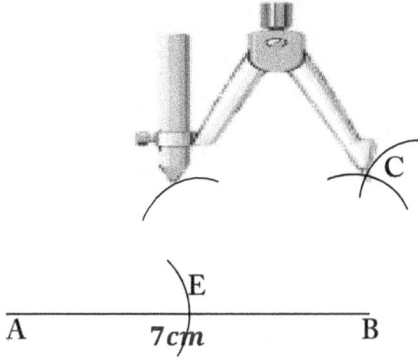

Step 6: With A as the centre and radius 5cm, draw another arc to cut the previous arc at D.

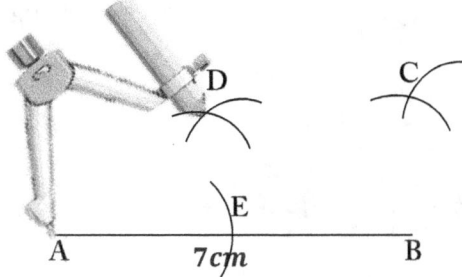

Step 7: Use ruler to join all the points together to form the required trapezium.

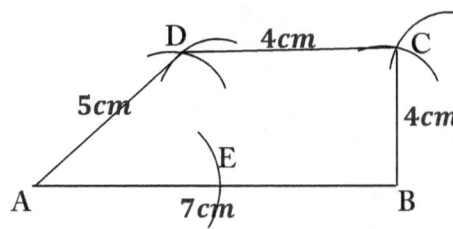

How to construct a trapezium when three sides and one angle is given

Example:

Construct a trapezium $PQRS$ in which $\overline{PQ}$ is parallel to $\overline{SR}$, $|PQ| = 8cm$, $\angle PQR = 75^0$, $|QR| = 6cm$, and $|PS| = 6cm$.

Procedures:

Step 1: Draw a line $|PQ| = 8cm$

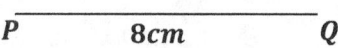

Step 2: With Q as centre, construct angle 75^0

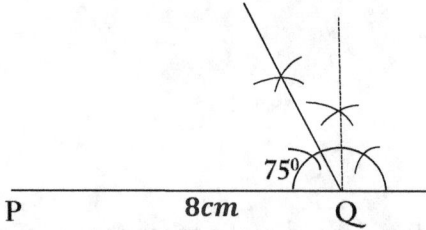

Step 3: With Q as centre and radius 6cm, draw an arc to cut the arm of the angle 75^0 at R

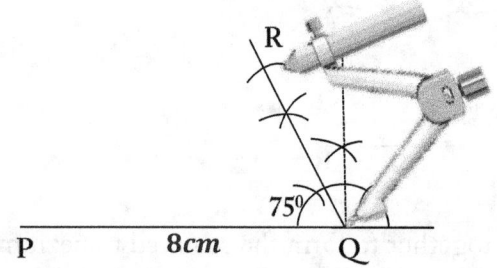

Step 4: Draw RX parallel to PQ

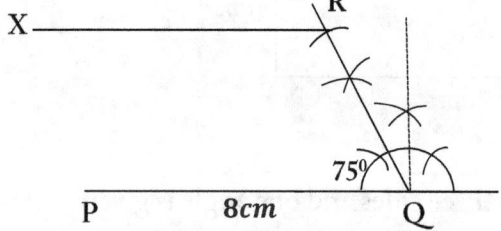

Step 5: With P as the centre and radius 6cm, draw an arc to cut the line RX at S.

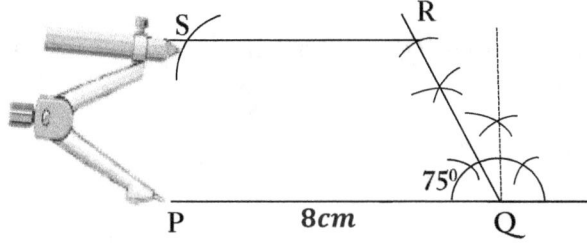

Step 6: Use ruler to join all the points together to form the required trapezium.

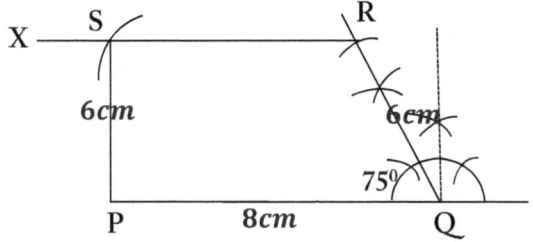

Exercise

1. Construct a parallelogram PQRS such that $|QS| = 9cm$ $|RS| = 6cm$ and $\angle QRS = 30^0$.
   Measure $|PR|$

----

2. Construct a square with sides 6cm long and measure the length of the diagonal.

3. Construct a rhombus with sides 6cm such that one of its acute angles is 60^0. Measure the diagonals of the rhombus.

----

4. Using a ruler and a pair of compasses only, construct a parallelogram PQRS such that $|PQ| = 8cm$, $|RQ| = 7cm$ and $\angle RQP = 135^0$. Measure $|QS|$

5. Using a ruler and a pair of compasses only, construct a trapezium ABCD such that $AB \parallel DC$ are 4cm apart, $\angle DAB = 60^0, |AB| = 8cm \ and \ |BC| = 5cm$. Measure $|DC|$

----

6. Construct a quadrilateral ABCD such that $|AB| = 8cm, |BC| = 6cm, \ |AC| = 10cm$. Measure $|BD|$ and write down the special name of ABCD.

7. Construct a parallelogram ABCD with $|AB| = 8cm$, $|AD| = 5cm$ and $\angle BAD = 75^0$. Measure the diagonals $|AC|$ and $|BD|$.

----

8. Construct a rectangle PQRS in which $|PQ| = 5cm$ and the diagonal $|PR| = 8cm$. Measure $|PS|$

CHAPTER SEVEN

Locus

Meaning of Locus

A locus in geometry is a set of points, whose location is satisfied or determined by one or more specified conditions for a shape or figure. Locus was derived from the Latin word meaning location or place. The plural of locus is "**loci**" and it is pronounced as "**losai**".

In other words, the set of the points that satisfy certain property is called the locus of a point. A locus of points can be a line, a curve or a surface.

Locus Theorem

The following are some common theorems on loci every student should know

Theorem 1

The locus of points at a given distance from another fixed point. *"This locus is a circle; whose radius is the given distance and the fixed point is the centre of the circle."*

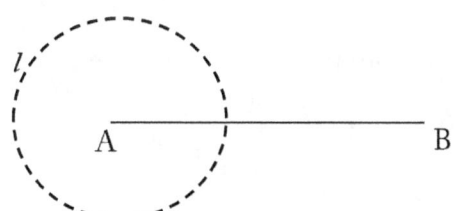

Theorem 2

The locus of points which is equidistant from two given fixed points. *"This locus is a perpendicular bisector of the line joining the two points together."*

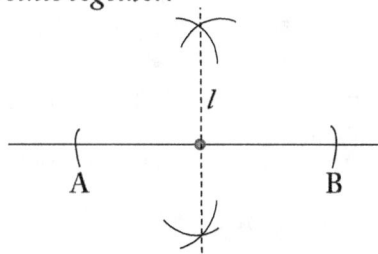

Theorem 3

The locus of points which is equidistant from two intersecting straight lines. *"This locus is the internal angle bisector of the two intersecting lines."*

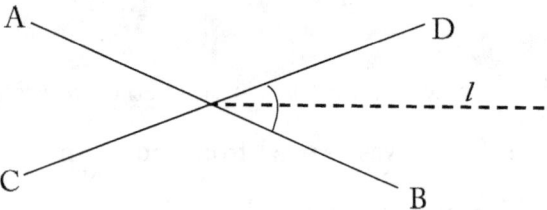

Theorem 4

The locus of points which is equidistant from a given distance of a given straight line. *"This locus is a set of parallel lines on opposite sides of the given straight line and the distance of the parallel lines from the given straight line is the same."*

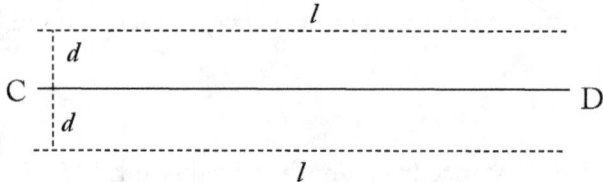

Theorem 5

The locus of point which is equidistant from two fixed parallel line. *"This locus is a line parallel to the two given straight line and at mid–way between them."*

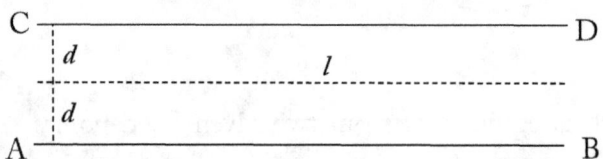

Example

Construct a $\triangle ABC$ such that $|AB| = 6cm$, $|AC| = 11cm$ and $A\hat{B}C = 135^0$. Construct the locus of l_1 points which is equidistant from A and B. Construct the locus l_2 of points which is 4.5cm from A. Locate the points P_1 and P_2 where l_1 and l_2 intersect. Measure $|P_1P_2|$

Procedures:

Step 1: Draw a line $|AB| = 6cm$

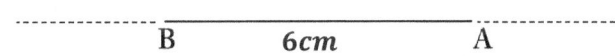

Step 2: With B as the centre construct angle 135^0

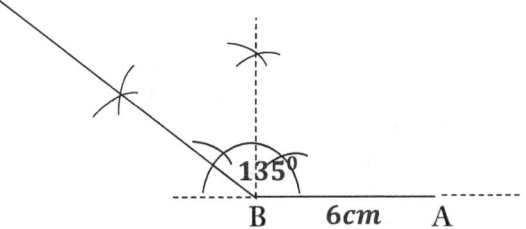

Step 3: With radius $11cm$ and at point A as the centre draw an arc to cut the arm of angle 135^0 and call the point of intersection C.

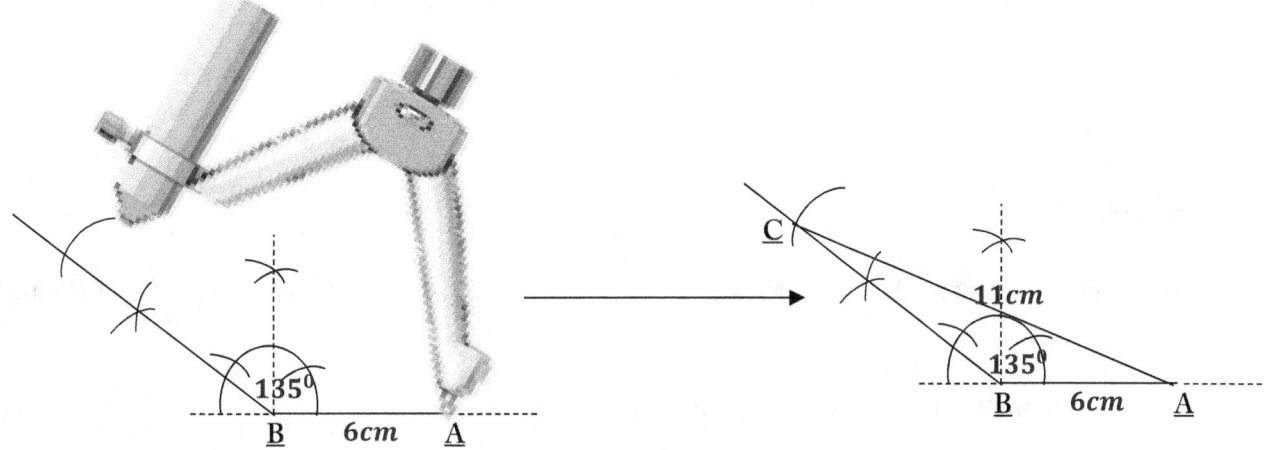

Step 4: Construct locus l_1, a perpendicular bisector to *A and B*

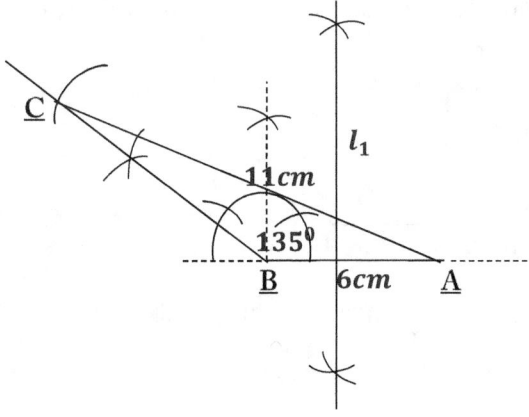

Step 5: With radius 4.5cm and point A as the centre, construct a circle i.e. locus l_2 to cut across locus l_1

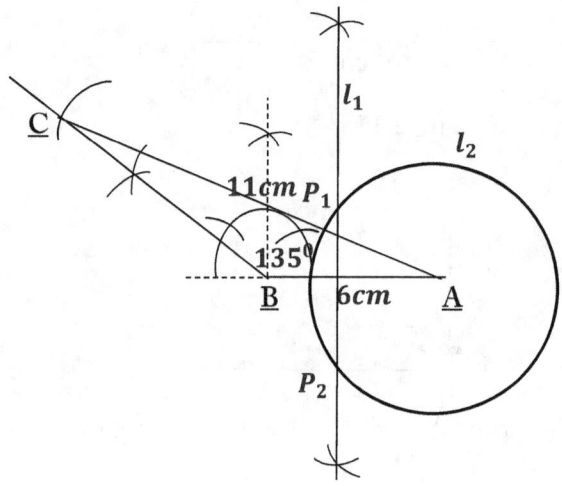

Step 6: Locate the points P_1 and P_2 where l_1 and l_2 intersect. Measure $|P_1P_2|$

$$|P_1P_2| = 6.7cm$$

Example

Construct a quadrilateral $PQRS$ such that $|PQ| = 5cm$, $|SR| = 3cm$, $|QR| = 6cm$. $QP \parallel RS$ and $|PR| = 9cm$.

(a) Draw the locus l_1 of points equidistant from $\overline{PS}$ *and* $\overline{PQ}$.

(b) Draw the locus l_2 of points equidistant from R and S.

(c) Given that T is the point of intersection of l_1 and l_2 and measure $|PT|$ and $|PS|$

Procedures:

Step 1: Draw a line $|PQ| = 5cm$

P $\quad$ 5cm $\quad$ Q

Step 2: With radius $6cm$ and Q as the centre draw an arc above

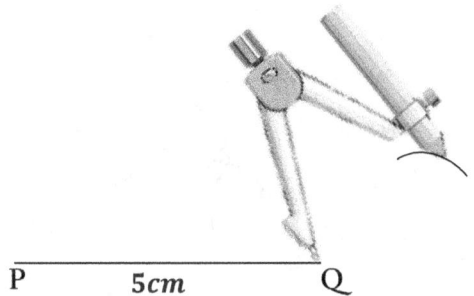

Step 3: With radius $9cm$ and P as the centre, draw an arc to cut the previous arc and call it R.

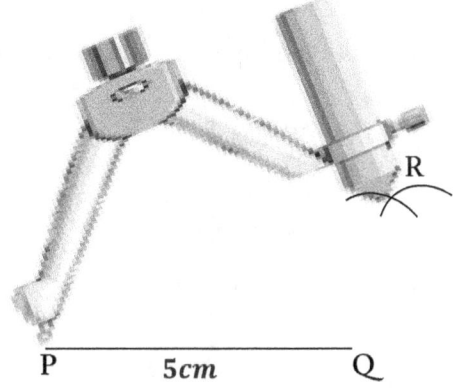

Step 4: With radius $3cm$ and R as the centre, draw an arc to cut the previous arc and call it S.

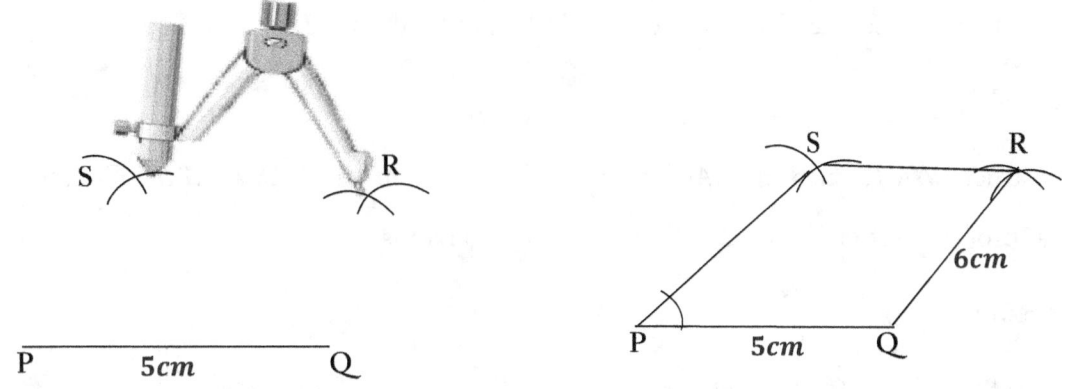

Step 5: Draw locus l_1 bisector of angle P.

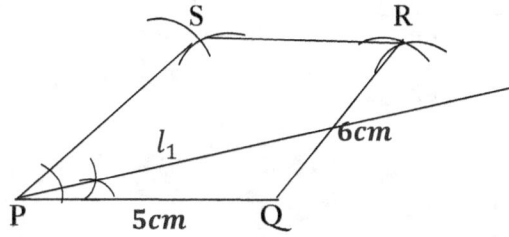

Step 6: Draw locus l_2 perpendicular bisector of $|RS|$

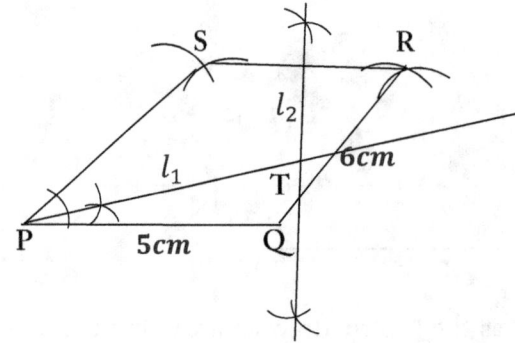

Step 7: Mark the point where l_1 and l_2 meet as T and measure and record the value of $|PT|$ and $|PS|$

$$|PT| = 6cm \text{ and } |PS| = 6.9cm$$

Circumcircle

A circumcircle is a circle whose circumference touches the three vertices of a triangle, thereby enclosing the triangle completely.

In constructing a circumcircle, all we need to do is bisect any two sides of the triangle and where both bisectors meet becomes the centre of the circle that touches the three vertices of the triangle.

Example

Construct a ΔABC such that $|AB| = 7cm$, $|BC| = 10cm$ and $B\hat{A}C = 105^0$. Construct a circle which pass through the vertices of ΔABC and measure its radius.

Procedures:

Step 1: Draw line $|AB| = 7cm$

-----------------------$\underline{\hspace{4cm}}$-----------------
$\qquad\qquad$ A $\qquad\qquad$ 7cm $\qquad\qquad$ B

Step 2: With A as the centre, construct angle 105^0

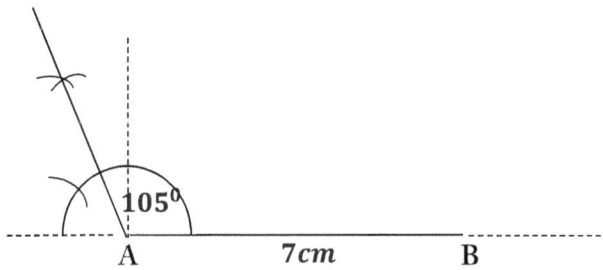

Step 3: With radius $10cm$ and B as the centre draw an arc to cut the arm of angle 105^0 and call it C.

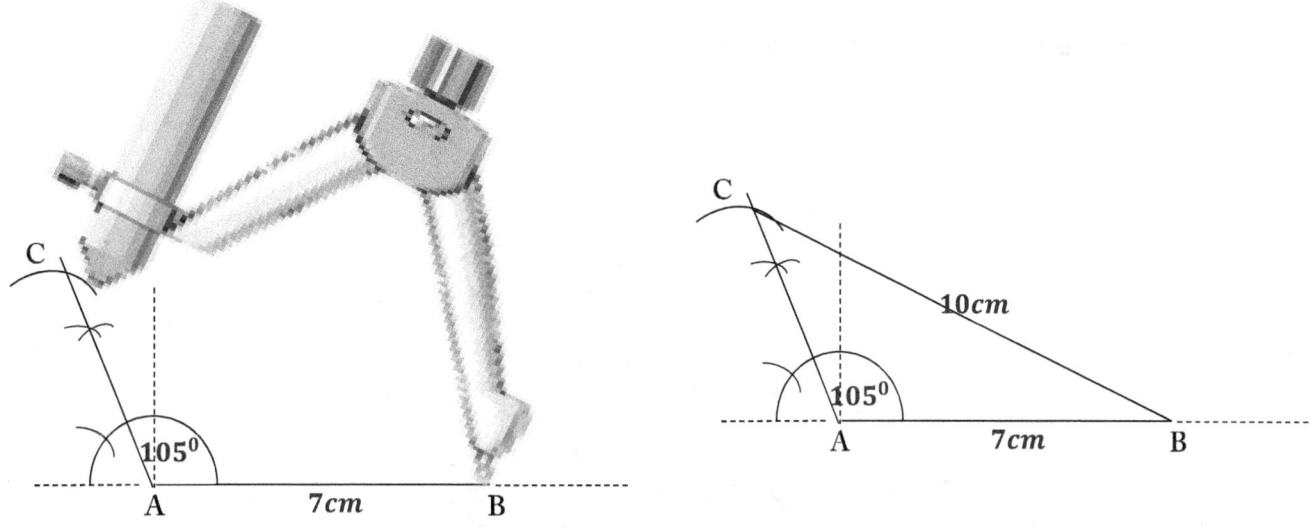

Step 4: Construct a perpendicular bisector of $|AB|$ and call it locus of l_1.

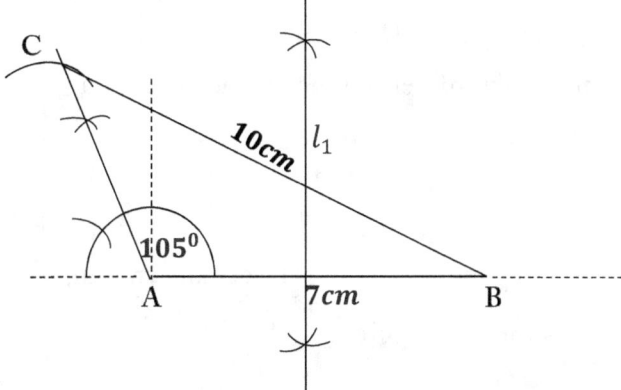

Step 5: Construct a perpendicular bisector of $|BC|$ and call it locus l_2.

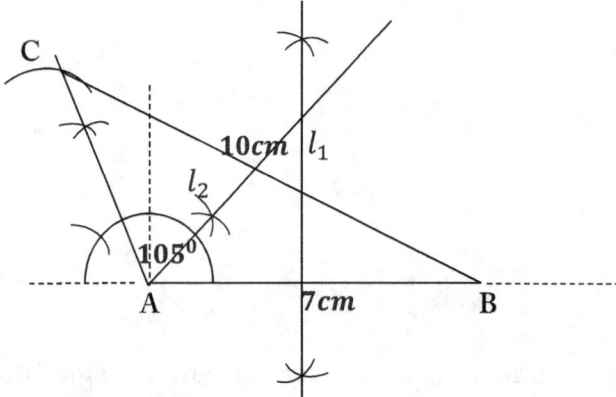

Step 6: Measure and record the radius of the circle.

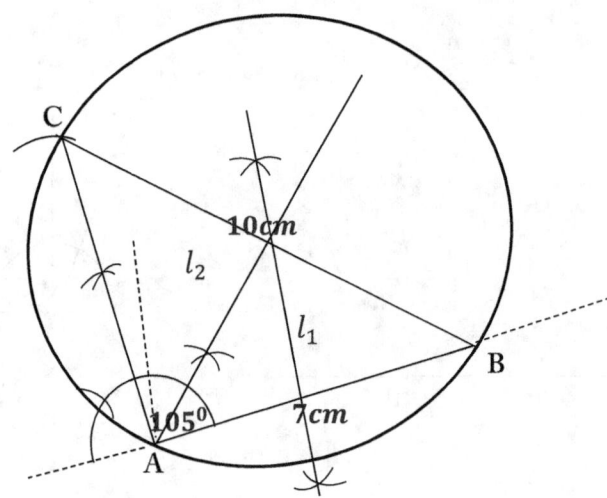

$$Radius = 5.1cm$$

Inscribed Circle

An inscribed circle is a circle whose circumference touches each side of a given triangle. An inscribed circle is also called *incircle*.

In constructing an inscribed circle, all you need to do is bisect any two angles of the triangle and where both bisectors meet becomes the centre of the circle.

Example

Construct a triangle ABC such that $|AB| = 5cm$, $|BC| = 6cm$ and $|AC| = 7cm$. Construct an inscribed circle.

Procedures:

Step 1: Draw line $|AB| = 5cm$

```
_____
A        5cm        B
```

Step 2: With radius $6cm$ and B as the centre, draw an arc above.

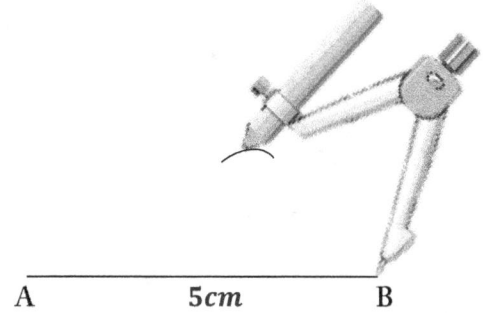

Step 3: With radius $7cm$ and A as the centre, draw an arc to cut the previous arc and call it C.

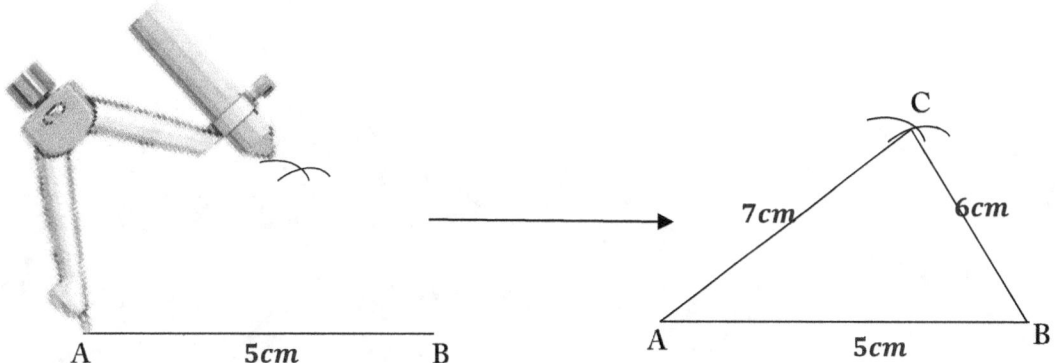

Step 4: Bisect angle A and call it locus of l_1

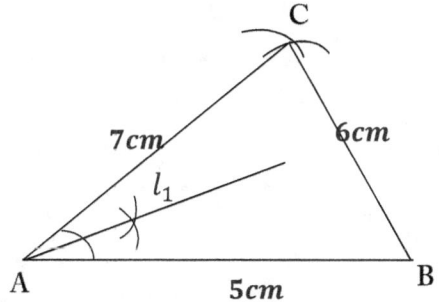

Step 5: Bisect angle B and call it locus of l_2. Where locus of l_1 and locus of l_2 meet becomes the centre of the circle.

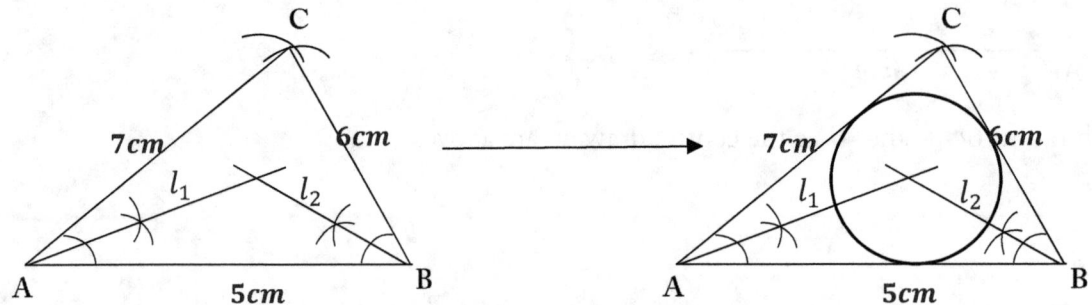

1. What is the size of the angle marked x in the diagram below? (a) 30^0 (b) 45^0 (c) 55^0 (d) 60^0 (e) 75^0

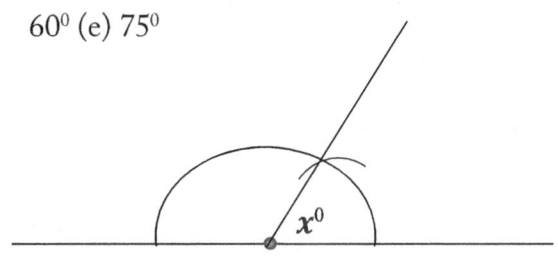

2. What is the size of the angle marked x in the diagram below? (a) 60^0 (b) 75^0 (c) 90^0 (d) 100^0 (e) 105^0

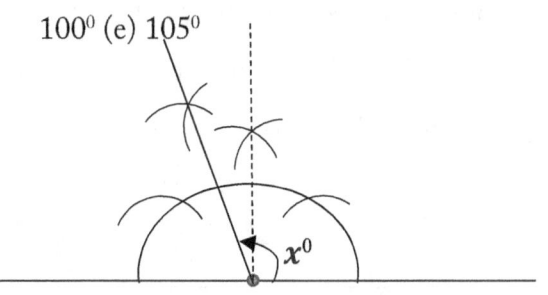

3. The angle marked k in the drawing below is _____ degrees. (a) 30^0 (b) 45^0 (c) 60^0 (d) 75^0 (e) 75^0

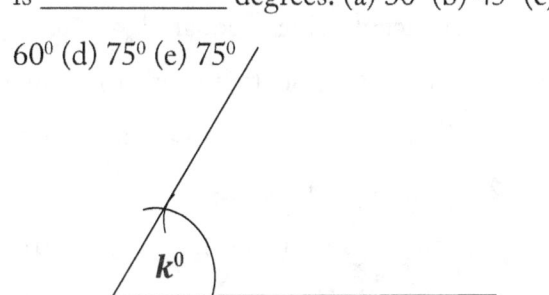

4. To construct angle 30^0, it is proper to construct angle (a) 30^0 (b) 45^0 (c) 60^0 (d) 75^0 (e) 90^0

5. The sum of angles on a straight line is _____ (a) 090^0 (b) 120^0 (c) 180^0 (d) 270^0 (e) 360^0

6. Which of the following angles cannot be constructed using a ruler and a pair of compasses only? (a) 30^0 (b) 45^0 (c) 60^0 (d) 15^0 (e) 70^0

7. Which of the following is not a material needed for the construction of an angle? (a) A pair of compasses (b) Plain paper (c) Protractor (d) Ruler (e) Scale pan

8. Two lines are perpendicular if the angle between them is _____ (a) 60^0 (b) 90^0 (c) 180^0 (d) 360^0 (e) 450^0

9. The bisection of angle 30^0 leads to angle. (a) 15^0 (b) 45^0 (c) 60^0 (d) 75^0 (e) 90^0

10. A triangle having two equal sides is known as _____ (a) equilateral (b) isosceles (c) obtuse (d) right angled (e) scalene

11. The instrument used for measuring, drawing and confirming angles is called _____ (a) a pair of compasses (b) a pair of dividers (c) the ruler (d) the protractor

12. What is the size of the angle ABD? (a) 155^0 (b) 135^0 (c) 75^0 (d) 45^0 (e) 90^0

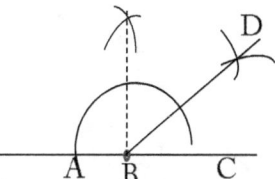

13. Which of the following is an obtuse angle? (a) 89^0 (b) 96^0 (c) 360^0 (d) 180^0 (e) 90^0

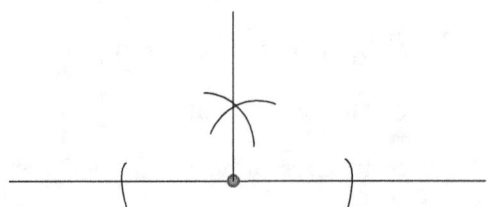

14. The construction above is that of angle _____ (a) 30^0 (b) 45^0 (c) 60^0 (d) 75^0 (e) 90^0

15. To construct angle 90^0, you will need to bisect angle. (a) 090^0 (b) 120^0 (c) 180^0 (d) 270^0 (e) 360^0

Use the construction below to answer question 16 and 17

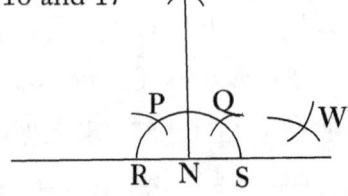

16. What is the acute angle formed when the point W is joined to N? (a) 20^0 (b) 45^0 (c) 30^0 (d) 25^0 (e) 90^0

17. What is the acute angle formed when ∠UNR is bisected? (a) 30^0 (b) 45^0 (c) 60^0 (d) 75^0 (e) 15^0

Use the construction below to answer question 18 to 20

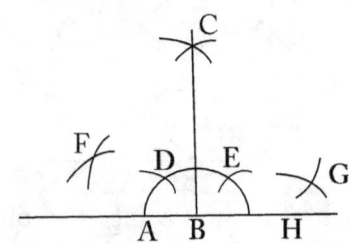

18. Find the acute angle GBH when point G is joined to B. (a) 30^0 (b) 20^0 (c) 60^0 (d) 75^0 (e) 15^0

19. What is the obtuse angle FBH when F is joined to B? (a) 120^0 (b) 105^0 (c) 100^0 (d) 135^0 (e) 45^0

20. Calculate the sum of ∠ABF and ∠GBH (a) 120^0 (b) 100^0 (c) 60^0 (d) 75^0 (e) 90^0

21. A triangle in which its angles are 110^0, 50^0 and 20^0 is an example of a/an _____ (a) acute angled (b) equilateral (c) Obtuse angled (d) isosceles (e) right angled

22. Which angle is constructed below? (a) 30^0 (b) 45^0 (c) 60^0 (d) 75^0 (e) 90^0

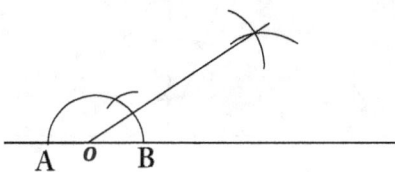

23. The name of the triangle with a pair of adjacent sides equal is called _____ triangle (a) acute angled (b) equilateral (c) Obtuse angled (d) isosceles (e) right angled

24. A triangle that has one of its angles to be more than 90^0 is called _____ triangle (a) acute angled (b) equilateral (c) Obtuse angled (d) isosceles (e) right angled

25. Which of the following pairs of plane figure have their diagonals equal and intersect at their centre? (a) kite and rectangle (b) parallelogram and rhombus (c)

parallelogram and kite (d) rectangle and square

26. Which of the following shows how to construct 45⁰? (a) I only (b) II only (c) III only (d) II and III only (e) I, II and III only

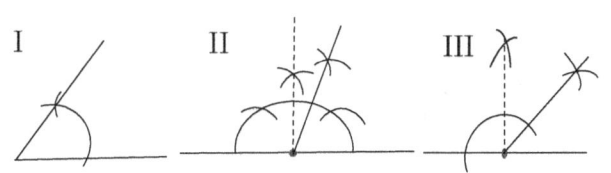

27. In bisecting any line, the following are required. (a) a ruler and a pair of dividers (b) a ruler and a protractor (c) a ruler and a set square (d) a ruler and a pair of compasses (e) a protractor and a pair of compasses.

28. Which of these is the correct sketch of angle 30⁰?   (a)   (b)

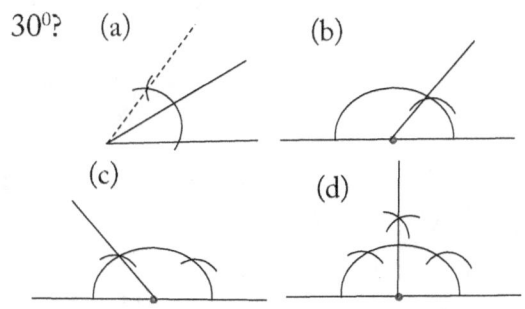

     (c)   (d)

29. A regular triangle is equilateral and a regular quadrilateral is a _____ (a) trapezium (b) square (c) rectangle (d) parallelogram (e) circle

30. Which of the following angle cannot be constructed using a ruler and a pair of

compasses only? (a) 30⁰ (b) 45⁰ (c) 55⁰ (d) 60⁰ (e) 90⁰

31. Which of the following method is/are appropriate for the construction of angle 60⁰?

(a) I only (b) II only (c) III only (d) II and III only (e) I and II only

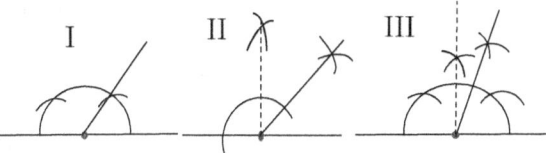

32. Which of the following is not a property of parallelogram? (a) all the sides are equal (b) each diagonal bisects the parallelogram into two congruent triangles (c) its opposite sides are parallel and equal (d) its opposite angles are equal (e) its diagonals bisect each other

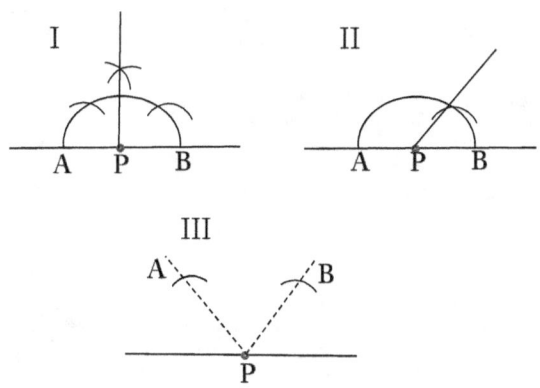

33. Which of the method above is/are the best way to construct a perpendicular to a line AB from point P? (a) I only (b) II only (c) III only (d) II and III only (e) I and II only

34. The angle bisector of an angle (a) divides the angle (b) divide the angle in two parts (c) divides the angle into two equal parts (d) none of these

35. A parallelogram PQRS is constructed with side QR = 6cm, PQ = 4cm and $\angle PQR = 90^0$. Then PQRS is a _____ (a) square (b) rectangle (c) rhombus (d) trapezium.

Answer Key to MCQ_1

1. D	2. E	3. C	4. C	5. C	6. E	7. E
8. B	9. A	10. B	11. D	12. D	13. B	14. E
15. C	16. C	17. B	18. A	19. D	20. D	21. B
22. A	23. D	24. C	25. D	26. C	27. D	28. A
29. B	30. C	31. A	32. A	33. A	34. C	35. B

Multiple Choice Questions 2

1. The point of intersection of the median of a triangle is called _____ (a) in-circle (b) circumcircle (c) orthocenter (d) centre

2. A point that is equidistant from the three sides of the triangle lies on the _____ (a) internal bisector of the angles of the triangle (b) perpendicular bisector of the triangle (c) bisector of the sides of the triangle (d) circumscribed circle of the triangle.

3. The perpendicular bisector of the sides of an acute-angled triangle are drawn. Which of these statements is correct? (a) on one of the vertices (b) at the midpoint of a side (c) inside the triangle (d) outside the triangle

4. *P* is a point on the same plane with a fixed point *A*. if *P* moves such that it is always equidistant from *A*, the locus of *P* is _____ (a) a straight line joining *P* and *A* (b) the perpendicular bisector of *AP* (c) a circle with centre *A* (d) the triangle with centre *P*

5. Which of the following statements describes the locus of a point R which moves in a plane such that it is equidistant from two intersecting lines? (a) the bisector of the angles formed by the lines (b) the point of intersection of the two lines (c) a cone with the two intersecting lines as slant height (d) a circle with the point of intersection of the two lines as the centre.

6. Which of the following statement is/are true? A. in a plane, the locus of points I. equidistant from a straight line is a circle with radius d where d is the distance between the point and the straight line. II. Equidistant from two given points *P* and *Q* is a circle of radius $|PQ|$. III. Equidistant from two points is the perpendicular bisector of the line joining the two points (a) I only (b) II only (c) III only (d) I, II and III only

7. The locus of all the points having a distance of *l units* from each of the two points **a** and **b** is: (a) a line parallel to the line **ab** (b) a line perpendicular to the line **ab** through the mid-point of **ab** (c) a circle through **a** and **b** with centre at the mid-point of **ab** (d) a circle with centre at a and passes through **b** (e) a circle in a plane perpendicular to **ab** and centre at the mid-point of the line **ab**.

8. P and Q are fixed points and X is a variable point which moves so that angle $PXQ = 45^0$. What is the locus of X? (a) a pair of straight line parallel to PQ (b) the perpendicular bisector of PQ (c) an arc of a circle passing through P and Q (d) a circle with diameter PQ (e) the bisector of angle PXQ

Use the diagram below to answer question 9 and 10

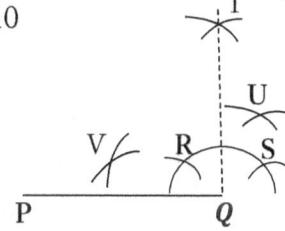

9. What is the obtuse angle formed when the point U is joined to Q? (a) 75^0 (b) 145^0 (c) 120^0 (d) 105^0 (e) 125^0

10. What is the acute angle formed when the point V is joined to Q? (a) 60^0 (b) 30^0 (c) 45^0 (d) 90^0 (e) 15^0

11. If U and V are two distinct fixed points and W is a variable point such that UWV is a right angle, what is the locus of W? (a) a perpendicular bisector of UV (b) a circle with UV as radius (c) a line parallel to the line UV (d) a circle with the line UV as diameter

12. What is the locus of the mid-point of all chords of length 6cm within a circle of

radius 5cm and with centre O? (a) a circle of radius 4cm and with centre O (b) the perpendicular bisector of the chords (c) a straight line passing through the centre O (d) a circle of radius 6cm and with centre O

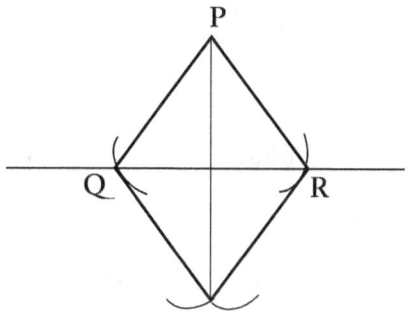

13. The figure above is an example of the construction of a (a) perpendicular bisector of a given straight line (b) perpendicular to the given point to a given line (c) perpendicular to a line from a given point on that line (d) given angle

14. PQR is a triangle in which $PQ = 10cm$ and $Q\hat{P}R = 60^0$. S is a point equidistant from P and Q. Also, S is a point equidistant from PQ and PR. If U is the foot of the perpendicular from S on PR, find the length SU in cm to one decimal place. (a) 2.7 (b) 2.9 (c) 3.1 (d) 3.3

15. The locus of point which moves so that it is equidistant from two intersecting straight line is the (a) perpendicular bisector of the

two lines (b) angle bisector of the two lines (c) bisector of the two lines (d) line parallel to the two lines

16. The locus of a point which is equidistant from two given fixed points is the (a) perpendicular bisector of the straight line joining them (b) parallel line to the straight line joining them (c) transverse to the straight line joining them (d) angle bisector of 90^0 which the straight line joining them makes with the horizonal.

17. What is the locus of a point P which moves on one side of a straight line XY, so that the angle XPY is always equal to 90^0? (a) perpendicular bisector of XY (b) a right-angled triangle (c) a circle (d) a semi-circle

18. If P and Q are fixed points and X is a point which moves so that $XP = XQ$, the locus of X is (a) a straight line (b) a circle (c) the bisector of angle PXQ (d) the perpendicular bisector of PQ.

19. Find the locus of a point which moves such that its distance from the line $y = 4$ is a constant, k (a) $y = 4 + k$ (b) $y = 4 - k$ (c) $y = k \pm 4$ (d) $y = 4 \pm k$

20. The locus of a point P which is equidistant from two given points S and T is (a) the

perpendicular bisector of ST (b) the angle bisector of PS and ST (c) a perpendicular to ST (d) a line parallel to ST

21. Which of these is the correct sketch of angle 30^0?

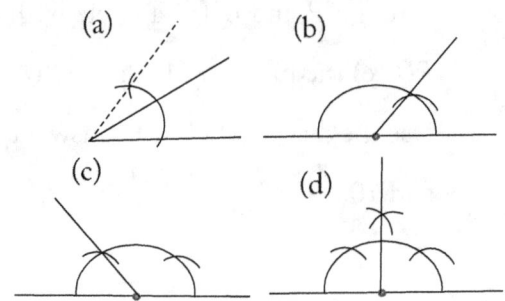

22. Determine the locus of a point inside a square $PQRS$ which is equidistant from PQ and QR. (a) the diagonal PR (b) the diagonal QS (c) side SR (d) the perpendicular bisector of PQ

23. The locus of a point which is 5cm from the line LM is a (a) a pair of lines on opposite sides of LM and parallel to it, each distance 5cm from LM (b) line parallel to LM and 5cm from LM (c) pair of parallel lines on one side LM and parallel to LM (d) line distance 10cm from LM and parallel to LM.

24. Two lines PQ and ST intersect at 75^0. The locus of the point equidistant from PQ and ST lies on the (a) perpendicular bisector of PQ (b) perpendicular bisector of ST (c)

bisector of the angle between the lines PQ and ST (d) bisector of the angles between lines PT and QS

25. If the locus of the points which are equidistant from P and Q meets line PQ at point N, then PN equals (a) NQ (b) $-NQ$ (c) $2NQ$ (d) $-2NQ$

26. A particle P moves between points S and T such that angle SPT is always constant. Find the locus of P. (a) it is a semi-circle with ST as diameter (b) it is a perpendicular bisector of ST (c) it is a quadrant of a circle with ST as a diameter (d) it is a straight line perpendicular to ST

27. What is the locus of the mid-point of all chords of length 6cm with a circle of radius 5cm and centre O? (a) A circle of radius 4cm with centre O (b) The perpendicular bisector of the chords (c) A straight line passing through centre O (d) A circle of radius 6cm and centre O

28. The perpendicular bisector of a line XT is the locus of a point (a) whose distance from X is always twice its distance from Y (b) whose distance from Y is always twice its distance from X (c) which moves on the line

XY (d) which is equidistant from the points X and Y

29. The locus of the points which is equidistant from the line PQ forms a (a) perpendicular line to PQ (b) circle with centre P (c) circle with centre Q (d) pair of parallel lines to PQ

30. The locus of a dog tethered to a pole with a rope of 4cm is a (a) semi-circle with radius 4m (b) circle with diameter 4m (c) circle with radius 4m (d) semi-circle with diameter 4m

31. Which of the following method is/are appropriate for the construction of angle 60^0?

(a) I only (b) II only (c) III only (d) II and III only (e) I and II only

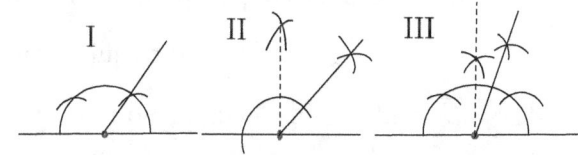

32. A point P moves so that it is equidistant from point L and M. If LM is 16cm, find the distance of P from LM when P is 10cm from L. (a) 12cm (b) 10cm (c) 8cm (d) 6cm

33. In constructing an angle, Olu draws line OX. With centre O and a convenient radius, he draws an arc interesting OX at P. With centre P and the same radius, he draws an

arc intersecting the first arc at Q and finally joins OQ. What is the size of angle POQ so constructed? (a) 90⁰ (b) 75⁰ (c) 60⁰ (d) 45⁰

34. The locus of points equidistant from two intersecting straight lines PQ and PR is (a) a circle centre P and radius Q (b) a circle centre P, radius PR (c) the point of intersection of the perpendicular bisectors of PQ and PR (d) the bisector of the angle QPR

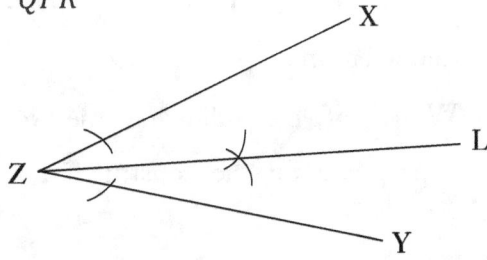

35. Describe the locus shown in the diagram below. (a) locus of point equidistant from X and Y (b) locus of points equidistant from X and Z (c) Locus of point equidistant from $\overline{ZX}$ and $\overline{ZY}$ (d) locus of points equidistant from $\overline{XY}$ and $\overline{ZY}$

36. Which of the following is NOT true? You may construct a (a) triangle given all the three sides (b) triangle given one side and two angles (c) triangle given two sides and the included angle (d) quadrilateral given all the four sides (e) parallelogram given two adjacent sides and the diagonal joining them

37. The locus of a point which is equidistant from two given fixed points is the (a) perpendicular bisector of the straight line joining them (b) angle bisector of the straight lines joining the points to the origin (c) perpendicular to the straight line joining them (d) parallel to the straight line joining them (e) a line making an acute angle with the line joining the two points.

Use the construction below to answer question 38 to 40

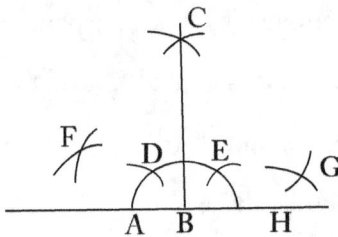

38. Find the acute angle GBH when point G is joined to B. (a) 30⁰ (b) 20⁰ (c) 60⁰ (d) 75⁰ (e) 15⁰

39. What is the obtuse angle FBH when F is joined to B? (a) 120⁰ (b) 105⁰ (c) 100⁰ (d) 135⁰ (e) 45⁰

40. Calculate the sum of $\angle ABF$ and $\angle GBH$ (a) 120⁰ (b) 100⁰ (c) 60⁰ (d) 75⁰ (e) 90⁰

1. Using a ruler and a pair of compasses only, construct: (a) $\triangle ABC$ such that $|AB| = 7.5cm$, $|AC| = 13.5cm$ and $\angle ABC = 120^0$ (b) the locus l_1 of point equidistant from point A and B. (c) the locus l_2 of points equidistant from B and C. (d) Using N, the point of intersection of l_1 and l_2 as the centre, draw a circle to pass through points A, B and C

2. Using a ruler and a pair of compasses only, construct: (a)i $\triangle ABC$ such that $|AB| = 5cm, |AC| = 13.5cm$ and $\angle CAB = 120^0$ (ii) the locus l_1 of point equidistant from point A and B (iii) the locus l_2 of points equidistant from $|AB|$ and $|AC|$, which passes through triangle ABC. (b) label the point P where l_1 and l_2 intersect. (c) Measure $|CP|$.

   Ans: $|CP| = 6.5cm$

3. Using a ruler and a pair of compasses only, construct a triangle ABC, given that $|AB| = 8.4cm, |BC| = 6.5cm$ and $\angle ABC = 120^0$. Construct the locus (a) locus l_1 of points equidistant from $|AB|$ and $|AC|$ and within the angle ABC (b) locus l_2 of points equidistant from B and C. locate the point of intersection P of l_1 and l_2, measure $|AP|$.

   Ans: $|AP| = 5.2cm$

4. $ABCD$ is a trapezium in which $AB \parallel DC, |AB| = 8cm, \angle ABC = 60^0, |BC| = 5.5cm$ and $|BD| = 8.3cm$ (a) Using a ruler and a pair of compasses only, construct (i) the trapezium $ABCD$ (ii) a rectangle $PQCD$, where P, Q are two points on AB. (b) Measure $|AC|$ and $|QB|$

   Ans: $|AC| = 7.2cm$ and $|QB| = 3.1cm$

5. Using a ruler and a pair of compasses only, construct (i) a triangle XYZ, in which $|YZ| = 8cm$, $\angle XYZ = 60^0$ and $XZY = 75^0$. Measure $|XY|$, (ii) the locus l_1 of point equidistant from point Y and Z; (iii) the locus l_2 of points equidistant from $|YX|$ and $|YZ|$. Measure QY where Q is the point of intersection of l_1 and l_2 Ans:
   $|XY| = 11.0cm, |QY| = 4.6cm$

6. Using a ruler and a pair of compasses only, construct a trapezium $ABCD$, in which the parallel sides AB and DC are 4cm apart. $\angle DAB = 60^0$ $|AB| = 8cm$ and $|BC| = 5cm$. Measure $|DC|$.

7. (a) Using a ruler and a pair of compasses only, construct $\triangle\, ABC$ in which $|AB\,| = 7cm, |BC\,| = 5cm$ and $\angle ABC = 75^0$. Measure AC (b) in (a) above, locate by construction, a point D such that CD is parallel to AB and D is equidistant from points A and C. Measure $\angle BAD$

Ans: $|AC| =$

7.5cm, $\angle BAD = 80^0$

8. Using a ruler and a pair of compasses only, construct (a) triangle QRT with $|QR\,| = 8cm, |RT\,| = 5cm\ and\ |QT\,| = 4.5cm$ (b) a quadrilateral $QRSP$ which has a common base QR with $\triangle\, QRT$ such that QTP is a straight line, $PQ\,\|\,SR, |QP| = 9cm\ and\ \ |RS| = 4.5cm$ (i) Measure $|PS|$ (ii) Find the perpendicular distance between RS and PQ. (iii) What is $QRSP$?

Ans: $|PS| = 5cm$ (ii) Perpendicular distance between RS and PQ is 4.7cm (iii) QRSP is Trapezium

9. Using a ruler and a pair of compasses only, construct (a) Construct a quadrilateral $MNOP$ with $|OP| = 11cm, |PM| = 6cm, |NO| = 5cm, |MN| = 8cm\ and\ |MO| = 9cm.$ (b) Measure $\angle MPO$. (c) Locate the point Y which is equidistant from the straight line ON and OP and also equidistant from points M and N (d) draw YP and find the perimeter of $\triangle\, OYP$.

Ans: $\angle MPO = 55^0$, the perimeter of $\triangle\, OYP =$

24.7cm

10. Using a ruler and a pair of compasses only, (a) Construct: (i) triangle XYZ $|XY| = 8cm, \angle YXZ = 60^0\ and\ \angle XYZ = 30^0$; (ii) the perpendicular ZT to meet XY in T; (iii) the locus l_1 of points equidistant from ZY and XY. (b) if l_1 and ZT intersect at S, measure $|ST|$

Ans: $|ST| =$

1.6cm

11. Using a ruler and a pair of compasses only, (a) construct a quadrilateral $PXYQ$ such that $|PX| = 9.9cm, |QX| = 10.2cm, \angle QPX = 75^0,\ |QY| = 10.4cm$ and $PQ\,\|\,XY$. (b) construct (i) the locus l_1 of point equidistant from point X and Y, (ii) the locus l_2 of points equidistant from $|QY|$ and $|YX|$. (c) locate M the point of intersection of l_1 and l_2 (d) Measure $|PM|$

12. Using a ruler and a pair of compasses only (a) construct: (i) $\triangle PQR$ such that $|PQ| = 8cm, |PR| = 7cm$ and $\angle QPR = 105^0$, (ii) the locus l_1 of point equidistant from point P and Q, (iii) the locus l_2 of points equidistant from point Q and R. (b) (i) Label the point T where l_1 and l_2 intersect. (ii) with centre T and radius $|TQ|$ construct a circle l_3. (iii) complete the quadrilateral $PQRS$ such that $|RS| = |QS|$ and $|TQ| = |TS|$

13. (a) Using a ruler and a pair of compasses only construct: (i) $\triangle XYZ$ such that $|XY| = 8cm$,and $\angle YXZ = \angle ZYX = 45^0$; (ii) locate a point P inside the triangle equidistant from XY and XZ, and also equidistant from YX and YZ; (iii) construct a circle touching the three sides of the triangle. (iv) measure the radius of the circle.

Ans: $radius = 1.7cm$

14. Using a ruler and a pair of compasses only (a) construct: (i) $\triangle PQR$ such that $|PQ| = 10cm, |QR| = 7cm$ and $\angle PQR = 90^0$, (ii) the locus l_1 of points equidistant from point Q and R, (iii) the locus l_2 of points equidistant from point P and Q. (b) Locate the point O equidistant from P, Q and R. (c) with O as the centre, draw the circumcircle of the triangle PQR. (d) Measure the radius of the circumcircle.

Ans: $radius = 6.2cm$

15. (a) Using a ruler and a pair of compasses only construct: (i) a quadrilateral $PQRS$ such that $|PQ| = 7cm, |PS| = 6.5cm$, $\angle QPS = 60^0, \angle PQR = 135^0$ and $QS \parallel QR$ (ii) the locus l_1 of points equidistant from point P and Q (iii) the locus l_2 of points equidistant from point P and S. (b)(i) Label the point T where l_1 and l_2 intersect. (ii) with centre T and radius $|TP|$, construct a circle l_3

16. (a) Using a ruler and a pair of compasses only construct: (i) a quadrilateral $PQRS$ such that $|PQ| = 7cm, |QR| = 8cm, |PS| = 6cm$, $\angle QPS = 75^0, \angle PQR = 60^0$ (ii) the locus l_1 of points equidistant from QR and RS; (iii) the locus l_2 of points equidistant from point R and S (b) Measure $|RS|$

Ans: $|RS| = 5cm$

17. Using a ruler and a pair of compasses only (a) construct a quadrilateral $PQRS$ such that $|PS| = 6cm, \angle RSP = 90^0 |RS| = 9cm, |QR| = 8.4cm$, and $|PQ| = 5.4cm$; (ii) the bisectors of $\angle RSP$ and $\angle SPQ$ to meet at X; (iii) the perpendicular XT to meet PS at T. (b) Measure $|XT|$

$$Ans: \qquad |XT| = 3.6cm$$

18. Using a ruler and a pair of compasses only, (a) construct a rhombus $PQRS$ of sides 7cm and $\angle PQR = 60^0$, (b) locate point X such that X lies on the locus of points equidistant from PQ and QR and also equidistant from Q and R; (c) Measure $|XR|$

$$Ans: \qquad |XR| = 4.0cm$$

19. (a) Using a ruler and a pair of compasses only, construct (i) a trapezium $WXYZ$ such that $|WX| = 10.2cm, |XY| = 5.6cm |YZ| = 5.8cm, \angle WXY = 60^0$ and $\overline{WX}$ is parallel to $\overline{YZ}$; (ii) a perpendicular from Z to meet WX at N. (b) Measure: (i) $|WZ|$ (ii) $|ZN|$

$$Ans: \quad |WZ| = 5.0cm, \quad |ZN| = 4.9cm$$

20. (a) Using a ruler and a pair of compasses only, construct (i) a trapezium $WXYZ$ such that $|WX| = 8cm, |XY| = 5.5cm |XZ| = 8.3cm, \angle WXY = 60^0$ and $\overline{WX} \parallel \overline{ZY}$; (ii) rectangle $PQYZ$ where P and Q are on WX. (b) Measure: (i) $|QX|$ (ii) $\angle XWZ$

$$Ans: (i) |QX| = 2.6cm (ii) \angle XWZ = 75^0$$

Answer Key to MCQ 2

1. B	2. C	3. C	4. C	5. A	6. C	7. B
8. B	9. D	10. B	11. D	12. A	13. A	14. D
15. B	16. A	17. D	18. D	19. D	20. A	21. A
22. B	23. A	24. C	25. A	26. A	27. A	28. D
29. A	30. C	31. A	32. D	33. C	34. D	35. C
36. D	37. A	38. A	39. D	40. D		

www.ingramcontent.com/pod-product-compliance
Lightning Source LLC
Chambersburg PA
CBHW081518220526
45467CB00010B/2962